C0-AXI-384

Tin: Its Production and Marketing

William Robertson

Contributions in Economics and Economic History, Number 51

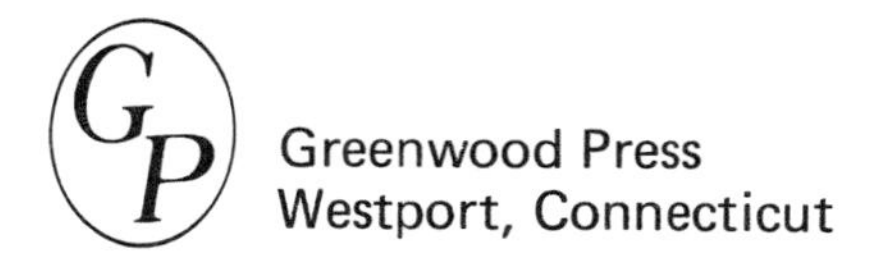

Greenwood Press
Westport, Connecticut

Published in the United States and Canada by
Greenwood Press, a division of Congressional
Information Service, Inc., Westport, Connecticut

English language edition, except the United States and Canada,
published by Croom Helm Ltd., London

JW

Library of Congress Cataloging in Publication Data

Robertson, William, 1920-
Tin, its production and marketing.

(Contributions in economics and economic history,
ISSN 0084-9235; no. 51)
Bibliography: p.204.
Includes index.
1. Tin industry. I. Title. II. Series.
HD9539.T5R625 1982 338.2'7453 82-9269
ISBN 0-313-23637-2 (lib. bdg.)

First published 1982

This series is edited by Fiona Gordon-Ashworth, formerly of the University of Southampton, who now works at the Bank of England. The views expressed in this book are not to be taken as those of the Bank of England.

Printed in Great Britain

CONTENTS

TABLES AND FIGURES

Tables

Figures

PREFACE

This book is a survey of the world tin industry, in which I have taken a keen interest for many years. For insight into the political economy of the industry I am much indebted to the late William Fox, the secretary to the International Tin Council from 1956 to 1971. I am very grateful to Bernard C. Engel, Deputy Buffer Stock Manager to the Tin Council, for reading the manuscript and for the helpful comments which he has made, but I am of course entirely responsible for the finished product. I have also received ready assistance from the statistical staff of the International Tin Council, who have patiently put up with my queries. For much of the material in this study I have found indispensable the statistical and other publications of the International Tin Council, and for a continuous picture of the world tin market I have relied heavily on the monthly journal *Tin International*. I am grateful to the University of Liverpool for study leave in which to complete the book, and to the secretarial staff of the Department of Economics for typing the manuscript.

NOTE

1. Changes have occurred in the names of several producing countries over the period covered in this study of the world tin market. Such countries are generally referred to here by their modern names: Thailand (Siam up to 1939), Indonesia (Netherlands East Indies up to 1945), Malaysia (the Federated and other Malay States up to 1963), Zaire (the Belgian Congo up to 1960 and the Democratic Republic of the Congo from 1960 to 1972). Tin production of the Congo up to 1942 included that of Rwanda.

2. Most of the data on consumption in this book refers to primary tin, namely, newly produced tin from the mines and smelters, but some secondary tin, in unknown amounts, is believed to be occasionally included in the International Tin Council's statistics, upon which the book chiefly relies. The consumption of secondary tin in the form or metal or otherwise is discussed in a separate chapter.

3. Tin concentrates are the material sent to the smelters for conversion into tin metal. Tin-in-concentrates measure the tin metal content of output from mines and producing countries.

INTRODUCTION

The high living standards of the industralized market economies and the prospects for improved living standards in the many developing countries at various stages of industrialization depend on the supply of metals. Over the last 100 years there has been an enormous growth in the consumption of metals, to produce which a great world mining industry has developed. Since Western Europe, Japan, and to some extent the US have outgrown their domestic production of metals, much of the output of the world mining industry enters into international trade. A substantial share comes from developing countries, whose domestic consumption of metals is still only a minute proportion of their output.

One of the most important non-ferrous metals in this world system of production and trade is tin. Unlike the other main non-ferrous metals, most tin deposits are found only in a small number of developing countries. The largest consuming country, the US, has no tin deposits, in marked contrast to a comparatively good endowment by nature with many other essential minerals. Since the Second World War, a new factor has entered into America's relationship with the tin mining industry, namely US ownership of a very large surplus stockpile. The problems arising from the disposal of the surplus have been a persistent source of argument between producers and the US government.

Tin has attracted a considerable amount of attention because it has been unique in being the subject of a continuous international commodity agreement for a quarter of a century. With so much argument over commodity market stabilization in the context of the debate on a new international economic order, tin has often been quoted as an example for other commodities, at least by spokesmen for the developing countries, not least because of its deemed suitability for buffer stock controls.

There is another unusual feature of the world tin market. Over many years the growth of consumption has been relatively low. Consumption of all the other important non-ferrous metals has grown faster since at least the late twenties. The relatively unfavourable growth rate for tin naturally raises questions about the key determinants of demand in both the past and the future.

Finally, as a non-renewable industrial mineral, tin has been considered by the resource pessimists to be sufficiently scarce as to threaten the world economy in the not too distant future with a serious supply problem.

The object of this book is to give a wide-ranging picture of the world tin market. It seems desirable that more should be done to disseminate knowledge about the factors affecting the market; the behaviour of production and consumption; trends and fluctuations in prices and costs; the role of foreign capital and technology in an industry with a substantial degree of state ownership and growing state participation in developing countries, and also with a considerable proportion of small-scale private production by nationals of developing countries, the problems of market stabilization, the adequacy of world supplies in the long run, and the problems of resource conservation. Drawing upon a number of disciplines, the book necessarily relies in some specialized fields on expert literature. Its main concern, however, is with the economics, or the political economy, of this relatively small, but important, section of the world commodity system.

1 A GENERAL REVIEW

The Nature of Tin Deposits

Tin-bearing ores are invariably associated with granitic rocks or the debris of such rocks.[1] The only ore of economic significance is cassiterite, containing in its purest form over 78 per cent tin. Cassiterite has a high specific gravity – 7.0 compared with 2.7 for most rock-forming minerals. This high specific gravity is exploited in separating cassiterite from the worthless materials associated with it. The cassiterite occurs either in thin, often very irregular veins or lodes, or in the debris which has built up from the gradual wearing-down of tin-bearing rocks in alluvial or eluvial deposits, which are found in river beds and valleys, or on the ocean floor close inshore. About 80 per cent of non-communist world tin production comes from these unconsolidated deposits, the rest from underground mines, chiefly in Bolivia. There are other underground mines in Cornwall, Australia, South Africa and Malaysia. Most Malaysian deposits, however, are alluvial, although it is possible that geological exploration may eventually lead to more underground mining in Malaysia in the distant future. The main areas of alluvial mining are in southeast Asia.

Compared with commercially viable deposits of copper, lead, zinc, nickel and bauxite, tin deposits are generally small. Few tin mines have exceeded an annual output of 2,000 tonnes even with modern methods. In contrast, copper, lead and zinc mines often produce over 100 tonnes a day. The minimum efficient scale of output for a modern Western-run copper mine would be 40,000 to 60,000 tonnes a year. There are certainly many small mines producing non-ferrous metals in developing countries, but in both developed and developing countries tin mining is exceptional in the comparatively small size of the most important mines, at least in terms of metal output, and the substantial proportion of world output which comes from a large number of extremely small units. An annual output of 30 tonnes is common in tin mining. Many mines have a very much smaller annual output, and several per cent of world output comes from purely individual operators working with the simplest equipment, often on a part-time basis.

The Geographical Distribution of Production

Out of 14 leading non-fuel minerals, tin comes at the top in terms of the proportion of world production coming from developing countries.[2] The proportion produced in the OECD countries is the second lowest after manganese.

As Table 1.1 shows, Malaysia is by far the largest producer, followed in 1979 by Thailand, which has recently replaced Bolivia in that position. These three countries accounted for about 63 per cent of non-communist production in 1978–9. Another 14 per cent comes from Indonesia (see Fig. 1.1). Two other developing countries, Nigeria and Zaire, have become much less important producers than they were 10 or 20 years ago, and have been surpassed by Brazil, the only new producing country to appear in the last 60 years. There are only three significant producers among the developed countries, Australia, the UK and South Africa, which together were the source of only 8.5 per cent of non-communist production in 1979. Both China and the Soviet Union are large producers, but official figures of their output are not available in the West.

Apart from these countries, the Tin Council records small quantities of tin production in another two dozen countries, the most important of which are Rwanda, Zimbabwe-Rhodesia, Peru, Burma, Laos and Spain. Only Rwanda currently produces more than 1,000 tonnes. Burma was an expanding producer during the interwar years, reaching its peak of over 5,000 tonnes in 1941, but in the postwar period has not exceeded 1,000 tonnes since 1955. Portugal and Spain were sizeable producers only in the special conditions of the war years. The world economy, therefore, depends essentially, as it always has done, on a very limited number of countries for most of its tin.

Trade and Consumption

The only important non-communist producing countries with a significant tin consumption are Australia, South Africa, the UK and Brazil. The other non-communist producing countries consume only about 1 per cent of their output, most of which they export to non-communist consuming countries. In 1977–8 tin accounted for about 2 per cent of the value of world primary commodity exports excluding petroleum, and came third in the list of important non-ferrous metals, as shown in Table 1.2.

Table 1.1: Production of Tin-in-Concentrates by Country[a] (000 tonnes)

	1929	1937	1950	1960	1970	1979
Malaysia	73.5	78.5	58.7	52.8	73.8	63.0
Thailand	10.1	16.1	10.5	12.3	21.8	34.0
Indonesia	37.4	38.2	32.6	23.0	19.1	29.4
Burma	2.6	5.2	1.5	0.9	0.3	0.8
Nigeria	11.3	11.0	8.4	7.8	8.0	2.8
Zaire[b]	1.0	9.1	11.9	9.2	6.5	3.3
South Africa	1.2	0.5	0.7	1.3	1.3	2.7
Bolivia	47.0	25.5	31.7	20.5	30.1	27.8
Brazil	–	–	0.2	1.6	3.7	6.6
UK	3.4	2.0	0.9	1.2	1.7	2.4
Spain	0.6	0.1	0.8	0.2	0.4	0.5
Australia	2.2	3.3	1.9	2.2	8.8	12.0
Other[c]	2 2.2	8.7	4.6	5.5	10.8	14.8
World (ex. communist countries)	192.5	198.2	164.4	138.5	186.3	200.9

Notes: a. All figures refer to tin metal content of tin-in-concentrates. b. Includes Rwanda in 1929 and 1937. c. Includes tin of unspecified origin, probably smuggled.

Source: ITC, *Report on the World Tin Position*, 1965. ITC, *Monthly Statistical Bulletin*, various issues. A. La Spada, *Statistics of Tin*, 1945-70.

Most tin exports go to a small number of industrial countries, of which the chief are, in order of current importance, the US, Japan, the Federal German Republic, the UK and France (see Table 1.3). Since the interwar and early postwar years the share of the US has been about halved, while the share of Japan has increased very substantially. The UK's share, at one time the largest, then second largest for most of this century, is now down to about 7 per cent, compared with Japan's 18 per cent, and a reduced 28 per cent share of the US (see Fig. 1.2). The UK's peak postwar consumption was as far back as 1947, its all-time peak was in 1928. In 1979 UK consumption was its lowest this century. As a percentage of total UK imports, tin is only a small item – 0.11 per cent in 1974. However, it is worth noting that the five main non-ferrous metals, tin, copper, lead, zinc and aluminium, collectively are only about 3 per cent of total UK imports. Their significance for the economy certainly cannot be judged by this small percentage.

The UK once had a large export trade in tin metal, importing concentrates for smelting chiefly from Bolivia and Nigeria, and exporting the metal to a large number of countries. In recent years imports of concentrates for smelting in the UK have greatly contracted, and exports of tin metal are now less than one-third of their level in the

Figure 1.1: Leading Tin Producers, 1945–1979

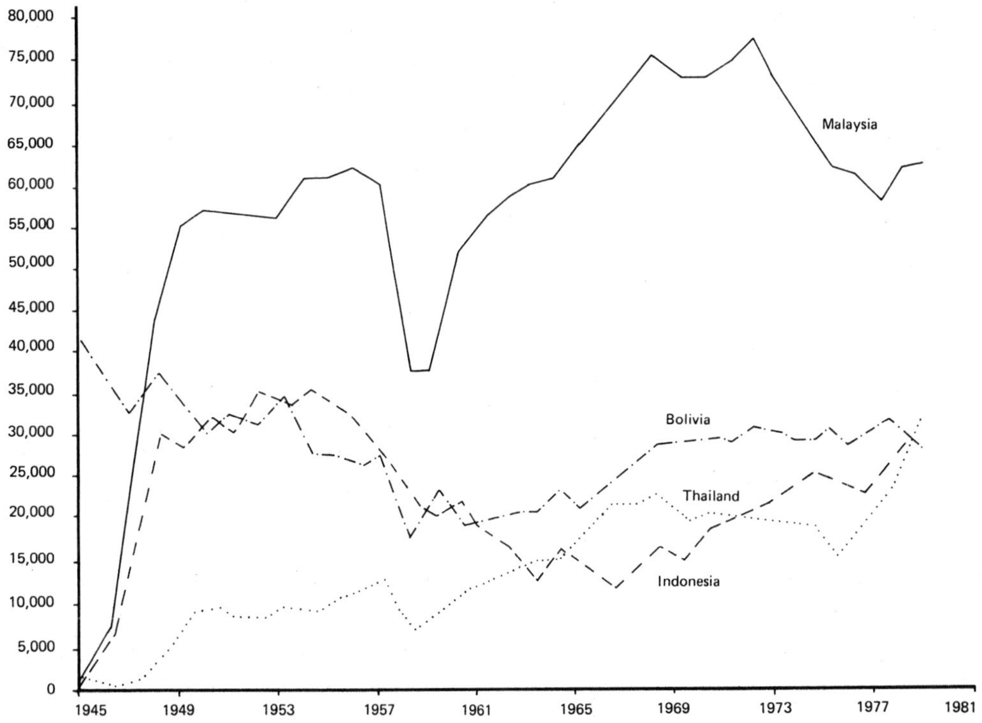

Table 1.2: World Market Economy Exports of Important Non-ferrous Metals, 1977-8 (million US dollars)

Copper (incl. ores)	9,413
Bauxite/aluminium	7,837
Tin	2,054
Zinc (incl. ores)	1,814
Nickel	1,491
Lead	1,237
Manganese ore	341
Tungsten ore	278

Source: UN, *Yearbook of International Trade Statistics*, 1979, Vol. 2, Trade by Commodity

Table 1.3: Consumption of Primary Tin Metal by Country (000 tonnes)

	1929	1937	1950	1960	1970	1979
USA	86.3	74.1	72.3	53.1	53.9	49.5
Canada	2.6	2.2	4.6	4.0	4.6	5.4
Japan	4.9	10.2	4.7	13.1	24.7	31.2
UK	24.6	26.4	23.7	22.1	17.0	11.1
German F.R.	16.3	13.4	7.9	(11.6)	14.1	13.7
France	11.9	9.3	7.9	11.4	10.5	10.0
Italy	5.0	3.7	3.9	5.1	7.2	6.0
Netherlands	1.1	2.9	3.1	3.2	5.5	4.8
Belgium	1.4	1.0	1.4	2.8	3.0	2.4
Australia	1.3	2.2	2.3	3.4	3.8	3.4
Switzerland	2.2	1.1	0.8	0.9	0.8	0.7
South Africa	0.7	0.6	1.2	2.0	2.1	2.0
Argentina	1.6	1.1	1.3	1.2	1.8	1.0
Brazil	0.7	0.9	1.6	1.6	2.5	6.0
Mexico	—	—	—	—	1.6	1.6
India	2.7	2.8	3.7	4.0	4.8	2.5
Other countries	10.1	10.2	9.5	25.5	16.3	20.0
World (ex. communist countries)	174.1	162.1	149.9	165.0	174.2	171.3

Source: ITC, *Report on the World Tin Position*, 1965. ITC, *Monthly Statistical Bulletin*, various issues. A. La Spada, *Tin Statistics*, 1945-70.

early fifties. Belgium formerly imported concentrates from the Congo, the Netherlands from the East Indies. Both Zaire and Indonesia now have their own smelters. Malaysia, though formerly a British dependency, did not export concentrates to the UK for smelting, since local smelters treated all Malaysia output, as well as a substantial part of the output of the rest of southeast Asia. Malaysian tin metal exports, therefore, always exceeded the output of the Malaysian mines. In the last

Figure 1.2: Leading Tin Consumers, 1945-1979

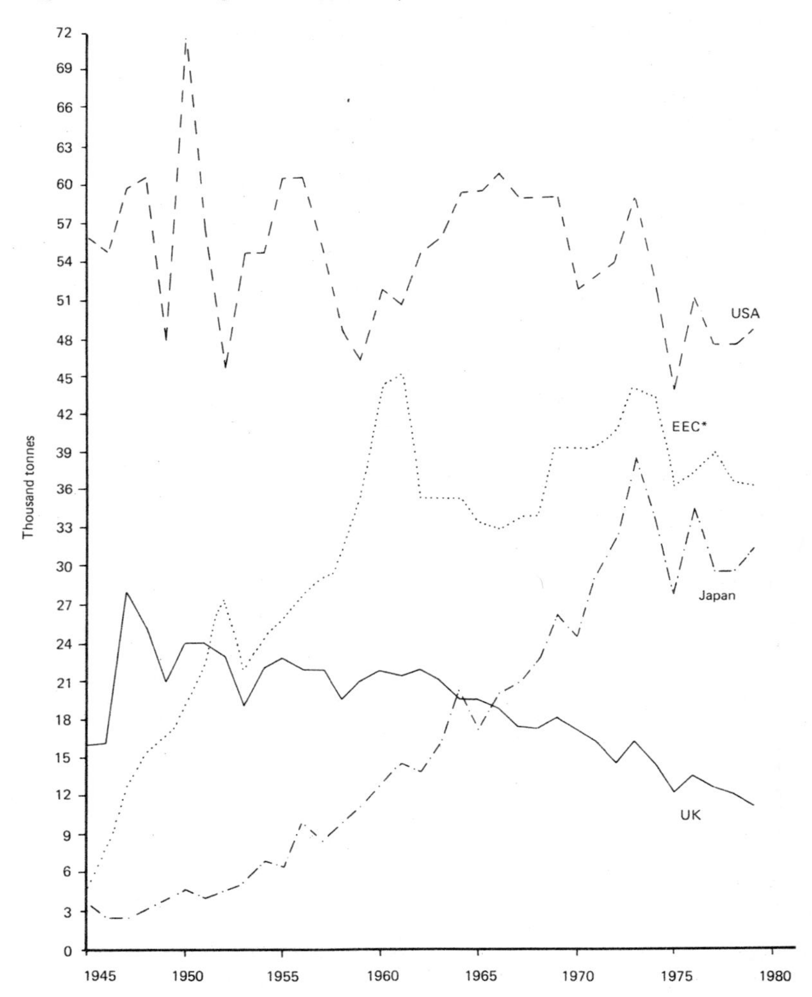

* The EEC excluding the UK

few years Malaysian exports have come closer to mine production, with the expansion of smelting capacity in countries which formerly sent concentrates to the Malaysian smelters.

Among the developing countries, Bolivia is by far the most dependent on its tin exports as a source of foreign currency and government tax revenue, but there has been a sharp fall in the share of tin in total Bolivian exports since the early fifties. From 83 per cent in 1950–2 it was down to 42 per cent in 1974–6, largely due to the growth of petroleum and natural gas exports. Tin has also become much less important in the export trade of the other five main producing developing countries. In Nigeria tin has been completely overshadowed by petroleum and now accounts for less than 1 per cent of total exports. Even in Malaysia there has been a considerable reduction in the share of tin, but it is still a significant item in exports and an important source of government revenue.

The Uses of Tin

Tin is one of the oldest metals known to man, its use in the very distant past confirmed by archaeological discoveries in various parts of the ancient civilized world.[3] Its position as an invaluable metal in a wide variety of uses has been established by reason of a number of properties: great malleability, low melting point, softness, corrosion resistance, non-toxicity, anti-friction qualities and appearance. In the present century, its main use has been in the coating of tinplate, currently accounting for about 40 per cent of non-communist world production. The second largest use is in solder, which is usually around one-quarter of consumption. The other main uses, bronze and brass, white metal, babbitt, anti-friction metal, tinning, and chemicals, account for much smaller percentages of world consumption (see Table 1.4). Tin is also used for a variety of lesser purposes, such as collapsible tubes and foil. At one time the coinage in parts of southeast Asia was made of tin.

Throughout the present century, world demand for tin has depended heavily on tinplate. In the late twenties, tinplate took about one-quarter of tin consumption, rising to one-third in the late thirties with the growing popularity of canned food. Since the Second World War, its share has fluctuated within the 40–45 per cent range, with a tendency in the late seventies towards the bottom end of the range.

In the familiar form of the tin can, tinplate has become one of the

Table 1.4: Tin Consumption by Uses[a] (tonnes)

	1965	1970	1975	1979
Tinplate	62,611	64,426	54,000	55,400
%	39.9	38.9	41.4	39.0
Solder	34,212	33,800	31,500	35,100
%	21.8	20.4	24.1	24.7
Bronze and brass	22,556	12,300	9,800	9,500
%	14.4	8.4	7.5	6.7
White metal, babbitt and anti-friction metal	11,016	12,100	12,000	12,500
%	7.0	8.3	9.2	8.8
Tinning	6,786	6,900	5,400	6,100
%	4.3	4.7	4.2	4.3
Other	19,816	16,300	17,800	23,500
%	12.6	11.2	13.6	16.5
Total[b]	156,997	146,100	130,500	142,100
%	100.0	100.0	100.0	100.0

Notes: a. Includes Brazil, Canada, USA, India, Japan, Austria, France, Germany FR, Italy, Switzerland, UK, Australia. b. Mainly primary tin, but some secondary tin metal is included (US tinplate only), UK (all uses).

Source: International Tin Council, *Tin Statistics*, 1965-75, 1969-79.

hallmarks of modern industrial civilization. It was a German invention, dating probably back to the beginning of the seventeenth century. After the development of the process of dipping iron plate into a vat of molten tin and draining off the surplus tin, a number of improvements were made in the eighteenth century, particularly by English producers who had started making tinplate in 1720. The invention of long-term food preservation in France, where Nicolas Appert discovered that contact with air was the chief cause of putrefaction, was followed by the application of the new processes for food preservation to containers made of tinplate. This happened in Britain and France around the same time towards the end of the Napoleonic wars. John Hall and Bryan Donkin, however, were probably the first to produce canned food in 1812. Appert was certainly doing so in 1815 at the 'House of Appert', the packaging company which he founded in 1812 on the basis of his successful invention.

Various improvements to tinplate were made during the rest of the century by British manufacturers who dominated the industry for many years. In 1856 a major step forward was achieved by the mechanization of the hot-dipping of tinplate. Much progress was also made in improving the quality and lowering the cost of the steel base of tinplate. The tin coating of hot-dipped tinplate even then was only a small percentage of the weight of the tinplate, the thickness of the coating being usually between 0.0001 and 0.0002 inch. No major change occurred in this respect until the 1940s, when the hot-dipping process began to be replaced by the electrolytic deposition of tin on the steel base, with a consequential reduction in the tin coating to as little as 15 or 30 millionths of an inch without any loss of quality in the tin-plate.

Over 90 per cent of tinplate is used for the manufacture of tin cans, a high proportion of which is used for the canning of food and beverages, with smaller proportions going into such things as paint cans, petrol cans, and the more modern product, aerosols. The non-container uses of tinplate include electric fires, reflectors, electronic equipment, toys and battery shells. There are considerable differences between countries, however, in the distribution of packaging of materials by product.[4] The proportion of tinplate used for food products and beverages is lower in Germany and France than in the US and the UK. Most beer and soft drinks containers in France are glass bottles, but in the UK, cans are preferred for beer and about half of soft drinks. Aluminium has a higher percentage of the beverage market in the US than in the UK.

The demand for tinplate in the last 30 years has been stimulated by the great expansion and diversity of the food canning industry. There is also a much more widespread consumption of canned foods than there was before the last war, particularly in continental Western Europe. Wider production of tinplate has been associated with this increase in consumption. Since the thirties, production outside the US and the UK has grown 12 times, and spread to a considerable number of less developed countries.

The growth of world tin can usage has made a very substantial contribution to the living standards of the mass of the people in all industrially advanced countries. As W.E. Minchinton has commented: 'As an expendable packaging for food and other products, tinplate (and hence tin) has been an important agent in the social revolution of the twentieth century by which the variety of diet has been increased and the labour of food preparation diminished. It has helped to provide great leisure in a world in which servants have disappeared'. Simply by the coating of steel plate with a very thin, but still vital layer of tin, it has been possible 'to preserve unlimited quantities of food in easily transportable form, to conserve food surpluses, to make available the resources of some parts of the world for those living in less favoured areas'.[5] It would be difficult to imagine modern Western civilization without tinplate. Although now there are alternatives competing for the vast packaging market, it is worth pointing out that only a few decades ago there were no aluminium cans, no plastic containers, no frozen foods or freezers. Apart from glass, it was only the tin can which served to keep many foods in an edible condition and in a convenient form over long periods. It was, in fact, and still is, one of the most sophisticated products of modern industrial civilization.

The second main use of tin is its combination with lead to form the tin alloys known as solders. The tin content of solder varies usually within the 15-65 per cent range, the proportions differing according to the purpose for which the solder is required. Some solders are used merely as fillers for automobile bodies, but to a diminishing extent. Others are used for making joints in the electrical and electronic industries. The oldest use is in plumbing.

There are many other tin-containing alloys. Bronze, an alloy of copper and up to 10 per cent tin, is probably the best known, being used in the manufacture of kitchen utensils, ornaments, musical instruments, marine bearings and brushes, and pumps. It has, of course, given its name to a period in human prehistory before 3000 BC – the Bronze Age. Bronze articles from Ur dating from 3500 BC have been found to contain 10–15 per cent tin. Bronze was certainly used in Egypt, China and Western Europe between 2500 and 1500 BC. In Peru, the Incas had their Bronze Age at the time of the Spanish invasion. Bronze articles, like tin in general, seem to have played an important part in international trade in the distant past.

Other alloys have been developed in recent times for a variety of purposes, tin-nickel alloys for printed circuits, tin-zinc alloys for hydraulic equipment, tin-aluminium alloys for bearings, cast-iron alloys with a small tin content for wear-resistant and heat-resistant irons.

About 4 per cent of tin is used for tinning a variety of goods, such as equipment for food processing and cooking, and for electrical and electronic components. In such products tin is invaluable because of its corrosion resistance, non-toxicity, appearance and fusibility. Tinning differs from tinplate in that the tin coating is applied directly to the steel-made article instead of to the steel sheets themselves.

The most recent substantial use of tin is in certain chemicals, namely a number of organotin and inorganotin compounds. The share of these compounds in world tin consumption is uncertain, but it is known to be an expanding source of demand, if not as buoyant a demand as was expected some years ago. Knorr estimated in 1944 that the prewar share of chemicals in US primary tin consumption was 1.8 per cent.[6]

In the late seventies it was probably about 10 per cent.

The first organotin compound was described by Löwig as far back as 1852. During the second half of last century and the first half of this century, there was a considerable amount of research into organotin chemistry, but it was not until the sixties and seventies that production of organotin compounds with a tin content ranging from 15 to 40 per cent began to expand rapidly. These compounds are used for the stabilization of PVC, for agricultural fungicides, industrial biocides and surface disinfectants.

Inorganotin compounds have a much longer history. According to J.W. Price, the Copts of Egypt are believed to have used a tin compound during the fifth century for the preparation of dyes.[7] The compound, stannic chloride, was first obtained as a liquid in 1605 by the alchemist Libavius. Stannic chloride and stannic oxide are estimated to account for about 80 per cent of the inorganotin compounds produced commercially. They are used for catalysts, colourants, veterinary medicines, toothpaste and soap.

A small recent use of tin is in molten form for the manufacture of float glass, but no other substantial use has emerged since the war. Some uses are now much less important than they were before the war; for example, tin sheets and tin foil have been largely replaced by aluminium. Tin tubes are still made in substantial quantities for various medical preparations where the chemical inertness of tin is important, but the amount of tin required for these purposes is only a small fraction of the thousands of tonnes once used. Similarly, technical change in the printing industry has virtually eliminated the demand for primary tin in type metal, and has also drastically reduced the demand for secondary tin in this use in the last ten years.[8]

The Pattern of Ownership

A feature of the changing structure of the world economy which is of great significance for mineral-importing industrial countries is the ownership and control of essential mineral resources. How the mining industry develops in mineral-rich exporting countries inevitably reflects the system of ownership and control. This is particularly important in less developed countries, where the mining industry has traditionally depended heavily, in many cases totally, on Western technology, capital and enterprise.

Important changes have occurred, and are still going on, in the structure of the world mining industry. In the Soviet Union and China tin mining is, of course, run by state enterprises. In the non-communist world there has been a very significant spread of state participation in the industry since the first wave of nationalization in the fifties.

One of the first countries outside the communist group to take over foreign tin mining enterprises was Burma, one of the smaller prewar producers. After a short-lived, unsuccessful experiment in joint ventures, the postwar independent Burmese government, which was basically opposed to foreign capital, nationalized the mining industry and put the administration of the country's mineral resources under a state mining development corporation. Bolivia nationalized about two-thirds of the industry in the 1952 revolution. Although a large part of the industry had been dominated by a Bolivian national, the well-known Simon Patiño, the company was a dollar corporation, located in Delaware, USA, and could essentially be regarded as a foreign enterprise.[9] Another part was owned by a company based in Switzerland. Their mines, and those owned by another of the so-called 'tin barons' of Bolivia, were nationalized under a new state corporation, COMIBOL (Corporación Minera de Bolivia). The rest of the industry is in the hands of a large number of privately-owned enterprises and co-operatives. Out of about 28 medium-sized firms, only two are foreign-controlled, another four or five have varying degrees of foreign minority participation. Other enterprises are generally very small, many of them extremely small units by any standard. Not much information is available about the co-operatives or the small mining units. The number of small producers apparently varies between 3,000 and 4,500, depending on economic conditions.[10]

Most Indonesian output comes from an entirely government-owned company which has overall responsibility for all tin mining in the country, and which took over the Dutch mining interests in the fifties. Several joint enterprises between the state corporation and private companies were set up during a period of relaxed conditions for foreign participation in tin mining after 1967. In 1979 about 15 per cent of Indonesian output came from a joint Australian-Indonesian enterprise. Another joint venture between the government company and the Dutch Billiton company, which formerly ran the Indonesian mines before the expropriation of Dutch interests, started producing tin in 1979. The only other foreign enterprise involved in the industry had not yet gone beyond the development stage at the end of the seventies.

As a result of postwar government takeovers in the three countries, about one-fifth of non-communist world tin production now comes from state mining companies. This does not measure the full extent of government intervention in the industry, which varies from country to country. In Malaysia a new government-owned organization, which was created in 1970, has recently acquired through a subsidiary a 70 per cent stake in a group of mining companies, which collectively account for about 30 per cent of Malaysian output.[11] In doing so, it also acquired through these companies a stake in Thailand's tin production. The other owner of the group is the large international mining finance house, Charter Consolidated, a company set up in 1965 with a wide range of commodity interests. There has also been an extension of public ownership in the Malaysian industry by the state governments as the result of dissatisfaction in the mining states of the Federation with their share of Malaysian income from tin. The Selangor government, for example, has entered into joint venture agreements with three mining companies, Pacific Tin Consolidated, the only US-controlled tin mining company in Malaysia, Brooklands Rubber Estate, the owner of recently-discovered tin-rich land, and the Berjuntai Tin Dredging Company, a member of the Malaysia Mining Corporation, in which the federal government's mining company, Permas, is involved with Charter Consolidated. The Selangor government also aims at a greater stake in all new and renewed mining leases in its territory.

An important strand in Malaysian mining policy during the seventies has been a commitment to give the indigenous Malays, as distinct from the Chinese citizens of Malaysia, a much greater stake in the industry . For many years the Malays have had only a very limited role in the mining industry, chiefly as workers in an unskilled capacity. At one time the Malays mined tin on a small scale, but their failure to overcome technical difficulties and mine more efficiently led to their replacement by immigrant Chinese miners, who succeeded where the Malays had failed. Until Western enterprise, supported by the colonial government, could beat Chinese techniques, organization and response to changing market conditions, Chinese miners dominated the industry after the decline of Malay mining in the middle of the nineteenth century.[12]

The modern Malaysian industry, therefore, has been built up, first by Chinese miners, and then by both Chinese and European miners. In the future it is intended by the government, in response to Malay nationalism, that a substantial Malay mining sector should be created. According to the federal government's economic plan, it is essential that the structure of ownership should reflect much more closely the distribution of the population by different ethnic groups. This policy implies a fall in both the Chinese and the foreign shares of productive capacity.

State participation has also begun in Thailand with the creation of a state-owned offshore mining organization in 1975. In 1975 the Thai government revoked the lease held by a major foreign company and specified that all new foreign dredging companies should have 51 per cent Thai ownership, rising to 60 per cent in five years. It also stated that the 60 per cent rule should be followed by existing companies wishing to renew their leases. It was reported in July 1980 that still tighter regulations were to be imposed on offshore mining. The local equity was to be raised from 60 per cent to 70 per cent within two years for companies which had completed 25-year mining concessions and five years for newer companies. An interesting outcome of the new Thai policy is that the partly state-owned Malaysian Mining Corporation has been required to reduce its holding in two of its constituent companies operating in Thailand.[13]

Tin production in Zaire is now concentrated in two companies, of which the larger, a merger of nine companies in the main tin-producing region, has 28 per cent state participation. Zairétain, the other company, is partly state-owned. In Nigeria, foreign investors are restricted to 40 per cent participation in mining enterprises.

Both Malaysia and Thailand have always had a large number of locally-owned enterprises, but not in the dredging sector. Their combined output has been a major proportion of the non-communist supply of tin for many years. Although there are Chinese-owned firms in Malaysia operating more than one mine, most are still small-scale partnerships or family businesses. Thailand also has many small producing units, most recently in the offshore part of the industry.

Summing up the position in the industry, therefore, changes have occurred in the following way since the Second World War. Among the main producing countries, other than Australia, the UK and South Africa, there is now a large public sector in two countries, a substantial, possibly growing, state interest in three countries, and an increase in private local participation in several developing countries. In all countries there is more state interference than before the war.

Even before the Malaysian government introduced its policy of promoting local Malay ownership, foreign investors were selling out in the early postwar years. Yip Yat Hoong points out that after the war local Chinese investors were increasingly interested in buying the tin shares of foreign-owned companies.[14] Possibly the unsettled political situation in the area was a factor encouraging this process of disinvestment by Western shareholders, but this can be only a partial explanation. Hoong suggests that there may have been some pessimism among Western investors about the industry's prospects. Whatever the full explanation, the result was substantial Chinese investment in the dredging sector of the Malaysian industry. According to Hoong, practically all the new shares issued by tin dredging companies between 1954 and 1964 were bought by local Chinese investors.

In contrast to other non-ferrous metals, tin did not attract until the late sixties the large international mining companies which are responsible for most of the non-communist world's output of minerals, and even then they made only a modest entry into the industry. This does not mean that there were no large companies in tin mining before the last war. On the contrary, there had been a large amount of integration in the interwar years, involving companies in Malaya, Nigeria, Cornwall, Bolivia and Thailand. Ownership links tied together mines in a number of countries and interlocking directorates were widespread. Moreover, mining engineers and mining consultants, common to many companies, had a powerful influence in promoting common policies and combinations in a large part of the prewar Malayan industry.[15] But all these companies were confined to tin, in contrast to multinational mining companies with interests in a number of minerals.

Notes

1. For a succinct account of the technology of tin mining see J. E. Denyer, 'The Production of Tin', a paper to the London Conference on Tin Consumption, 1972, upon which this section has relied for technical information. The paper is published in the *Conference Proceedings*, International Tin Council, London, 1972, pp. 47-54.

2. The commodities are bauxite, chromium, cobalt, copper, iron ore, manganese, mercury, lead, molybdenum, nickel, the platinum group, tin, tungsten and zinc. Estimated world production includes unofficial figures for the communist countries. See R.F. Mikesell, *New Patterns of World Mineral Development*, British-North American Committee, London, 1979.

3. For a study of tin's various uses and its role in social and economic history over many centuries see E.S. Hedges, *Tin in Social and Economic History*, Edward Arnold, London, 1964. This section draws heavily on Hedges' study.

4. *Tin International*, January 1980.

5. Walter E. Minchinton, *The British Tinplate Industry*, Clarendon Press, Oxford, 1957, p. 248.

6. K.E. Knorr, *Tin under Control*, Food Research Institute, Stanford University, Stanford, Calif., 1945, p. 40. Including secondary tin, the share of chemicals in total US tin consumption was put at 3.5 per cent on average over the 1935-9 period by Knorr.

7. J.W. Price, 'Inorganic Tin Compounds', a paper to the London Conference

on Tin Consumption, 1972, *Proceedings*, p. 201.

8. Only US figures for the use of tin in type metal are available. These show that less than 100 tonnes of primary tin have been used annually for type metal for many years, and as little as 26 tonnes in 1979. Up to the late sixties, over 1,000 tonnes of secondary tin or recycled tin was used in recycled type metal, but the annual rate of consumption was falling. Since 1969 there has been a drastic reduction to less than 150 tonnes in the last three years.

9. For a good account of the Bolivian situation, see William Fox, *Tin: the Working of a Commodity Agreement*, Mining Journal Books, London, 1974, pp. 57–69.

10. Malcolm Gillis *et al., Taxation and Mining: Non-Fuel Minerals in Bolivia and Other Countries*, Ballinger, Cam., Mass., 1978, p. 32.

11. Recent changes in the structure of the Malaysian industry are described in *Tin International*, various issues.

12. For the early history of tin mining in Malaysia see Yip Yat Hoong, *The Development of the Tin Mining Industry in Malaya*, University of Malaya Press, Kuala Lumpur, 1969. See also the article by Anthony Smith in *Tin International*, May, 1981.

13. ITC, *Notes on Tin*, June 1980, quoting *Business Times*, June 30 1980.

14. Yip Yat Hoong, op. cit., p. 365.

15. J.W.F. Rowe, *Primary Commodities in International Trade*, Cam. Univ. Press, London, 1965, p. 16.

2 PRODUCTION METHODS AND COSTS

Most of the world's tin is mined in one of three very different ways: by dredging, either onshore or offshore; by gravel pumping; by hard-rock underground mining. Collectively, these three mining methods account for at least 85 per cent of world output (see Table 2.1).

Dredging

Dredges are used for the mining of alluvial or eluvial tin deposits. A dredge is a kind of mechanized floating factory which moves along an artificially-created lake, excavating the tin-bearing ground as it goes along, and thereby extending the lake it has created. The over-burden has first to be removed to get at the pay-dirt at the bottom of the lake. A chain of buckets carries the tin-bearing ground on to the dredge where the first separation of the tin ore and the waste product takes place. The waste product, still containing some tin ore, is pumped over the back of the dredge. The capacity of a dredge is measured by the 'yardage' removed, that is, so many cubic yards of tin-bearing ground per unit of time, say a month. The output of the dredge depends on the size of the buckets and the speed at which they move.

The first dredge in tin mining was used offshore in Thailand as early as 1907. The onshore dredge was introduced into British Malaya in 1912. On the first Malayan dredge 10 cubic ft. buckets ran at a rate of 11 a minute, dug to a maximum depth of 50 feet and had an average monthly throughout of 57,000 cubic yards. Since then, the capacity of dredges has increased enormously. A typical modern dredge would have 24 cubic ft. buckets, running at perhaps 36 buckets a minute on a close connected band, with a monthly throughout of about 850,000 cu. yds.

Since the first bucket dredge was introduced into pre-1914 Malaya the basic design has not changed much, but the modern dredge incorporates many advances in engineering and electical technology, in metallurgy and mineral processing, which have helped to check the rise in costs as lower grades of ground are worked. With no change in the basic design 'the tendency has been to scale up what has served before, as the demand for increased capacity has risen, a practice which works

Table 2.1: Tin Production by Method of Mining, January-June 1980

	Number of units[a]	Production of tin-in-concentrates (tonnes)	Share of world production (percentage)
Dredges			
Offshore	31	7,332	7.3
Onshore	87	12,534	12.5
Suction boats	—	8,500	8.4
Gravel pumps	1,430	28,815	28.5
Underground mines	34	20,184	19.9
Opencast	28	1,918	1.9
Dulang washing	—	2,100	2.1
Other, n.e.s.[b]	—	19,517	19.4
Total		100,900	100.0

Notes: a. Refers to number of dredges, mines, gravel pumps, etc., in operation at end of June 1980. b. Included under 'Other' are figures for some output excluded by the reporting countries, also figures for Nigeria, Zaire, Brazil, Rwanda, Zimbabwe-Rhodesia, Burma, and a number of other smaller producing countries. The Soviet Union and China are excluded.

Source: ITC, *Monthly Statistical Bulletin*, January 1981.

satisfactorily only up to a certain limiting capacity'.[1] A disproportionate increase in operating costs results from the continued scaling-up in size of the bucket and increase in bucket speed.

A growing proportion of tin is mined by offshore dredges, embodying the same technology as onshore dredges, but so far, offshore dredges can operate only up to a limited depth in relatively calm waters close to the shore. The latest offshore dredges are very large mining units with a capacity of up to 1,500 tonnes of tin-in-concentrates. On the other hand, also operating offshore are large numbers of small craft, using the suction method of raising tin-bearing alluvium from the sea bed, averaging only about 3–4 tonnes. These suction boats have been working in Thai waters during the seventies, first illegally, then with government approval. Most of their output comes in the first and last quarter of each year, outside the monsoon season. Their method of operation is described in a Tin Council report.

> In their initial conception the suction boats provided a method of stealing ore from seabed deposits in leases belonging to the established dredging companies. The boats are fitted with a suction pipe connected to a gravel pump and the placing of the end of the suctio suction pipe on the seabed is controlled by a diver, often operating with makeshift diving equipment. The loose sand and gravel from

the seabed is sucked up through the suction pipe and discharged over a very small palong built over the deck of the boat, which recovers about 30 per cent of the available tin, with the rest being washed overboard with the tailings.[2]

Gravel Pumps and Other Methods of Surface Mining

Currently the most important method of mining tin in southeast Asia is by gravel pump. The gravel pump is a highly effective mechanism for dealing with smaller, or more difficult, alluvial deposits. In Malaysia, for example, the alluvial tin deposits occur on an underlying bedrock usually of hard limestone with a very uneven profile. The cassiterite ore tends to be located among limestone pinnacles which can be damaging to dredges, and make it impossible to extract all the ore by dredging. Gravel pumps, however, using powerful jets of water, can be used to disintegrate the tin-bearing alluvium, which is then pumped into a sluice where the cassiterite is tapped, and the lighter, waste material poured away.

It follows from the nature of these deposits that gravel pumps can work ground with which dredges cannot cope, and often take over where dredges have left off as uneconomic. Gravel pumps can also work smaller alluvial deposits for which a dredge, with its much higher capacity, could not economically be used. Since many tin deposits are, in fact, small and scattered, there is great scope for gravel pump mining. Not all gravel pump mines, however, are small. Some large Malaysian gravel pump mines are capable of producing up to 1,000 tonnes of tin a year, over 20 times greater than the average for Malaysia.[3]

In several countries, lode or placer deposits, resonably close to the surface, are worked by opencast mines using earthmoving equipment of various kinds. Most opencast mines in Malaysia are small. There are several large opencast mines in Australia, Nigeria, Zaire, and one in Malaysia. The largest opencast mine in Nigeria produced over 2,000 tonnes in 1978.

A small amount of tin is produced by a method known as hydraulicing. Like gravel pumping, it also involves the extensive use of water, but it differs from gravel pumping in that a natural head of water is used instead of power-propelled jets.

Each alluvial mining method produces tin concentrates containing waste material. On the site, sand is separated by gravity separation techniques. Magnetic and electrostatic separators remove the monazite, ilemite and zircon. Flotation techniques dispose of the sulphides. The cassiterite ore is then upgraded to the point where it exceeds 70 per cent Sn., after which it goes as tin concentrates to the smelter.

A form of mineral extraction peculiar to tin mining among the nonferrous minerals is dulang washing, so-called from the dulang or pan which is used to recover the cassiterite. Dulang washing is simply the recovery of tin ore from mine tailings and river beds by means of the most elementary equipment, similar to the method used by the old-time gold miners of North America and Australia. The amounts of tin recovered by the dulang washers, generally women, are individually very small, but collectively this ultra-small-scale operation adds up to about 5 per cent of Malaysian output and 2-3 per cent of Thailand's output.

Underground Mining

Underground mining of tin is generally deep mining, since most known near-surface deposits suitable for underground mining have been exhausted. Underground mining is less important than it was at the beginning of the century, before dredging had started in southeast Asia. Currently, the proportion of non-communist world output from underground mines is about 20 per cent most of it from Bolivia.

The size of underground mines varies enormously. They can be highly complex underground workings, at great depths, with many miles of tunnels, using extensive mining machinery and large quantities of explosives. On the other hand, many underground mines are small operations, often smaller than the typical gravel pump mines of southeast Asia. In Bolivia there are several thousand of such small units co-existing with huge underground mines, often in close proximity. Mining methods used in these small units are simple and often dangerous. As one Western expert has put it: 'In effect, too often life is cheaper than equipment. To equip a small shaft properly will cost many thousands of dollars. To dispose of a body, probably less than a hundred dollars will be required'.[4]

The Economics of Different Mining Methods

The Scale of Production

As the previous section has indicated, there is a very wide range between the largest and the smallest mining units, even if the dulang washers and suction boat operators are excluded. At one end are the large Bolivian and Australian mines, at the other the small gravel pump mines. The largest mine currently is the Renison complex in Tasmania, with an output in excess of 5,000 tonnes. Average output of the large Bolivian mines is over 1,000 tonnes. The one major Malaysian underground mine produces about 1,200 tonnes. The UK mines average about 800 tonnes.

Average output of dredges in southeast Asia increased substantially between 1921 and 1940. For some 30 Malaysian dredges in 1921 the average output was 167 tonnes. By 1940 the number of Malaysia dredges had increased to over 100, and their average output in that boom year, when wartime demand was virtually unlimited, had risen to 400 tonnes. In the early postwar years a major programme of dredge rehabilitation had to be carried out in both Malaysia and Indonesia to repair the damage and neglect of the war years. Between 1949 and 1958 36 Malaysian dredges were modernized, another 14 between 1960 and 1964. A similar reconstruction programme was undertaken by the Dutch before the independent Indonesian government took over the mines.

There has been no further increase in the average output of the Malaysian dredges, but new dredges of larger capacity are expected to recover in excess of 1,000 tonnes of tin-in-concentrates annually. Unlike the Malaysian industry, the Indonesian includes both onshore and offshore dredges. The average output of the former is about 200 tonnes, of the latter over 700 tonnes, with the newest additions to the offshore dredging fleet reaching 1,500 tonnes. Thailand also has both onshore and offshore dredges. Over the last ten years, depending on the number and capacity of dredges in operation, average output has varied between 240 and 300 tonnes.

Average output of the Malaysian gravel pump mines in 1978 was 40 tonnes, compared with 30 tonnes in Thailand and about 70 tonnes in Indonesia. The average has not changed since the fifties, whereas in Thailand there has been a fall from around 45-50 to 30 tonnes, with a large increase in the number of relatively small gravel pump mines. Until 1961 there were less than 100 such mines; at the end of 1979, 386 were recorded in operation. Indonesia had 177 gravel pump mines at the end of 1979 with an average output of about 70 tonnes.

Labour Productivity

Throughout the world tin mining industry there are very great differences in labour productivity. These differences are due to various factors: geological conditions, methods of mining, some of which are more capital-intensive and energy-intensive than others, recovery rates in treatment plants at the mines, managerial and technical efficiency.

Changes in mining methods and various technical innovations over the last 70 years have had a massive effect on employment in the industry, particularly in Malaysia. At the beginning of this century about 200,000 workers are estimated to have been employed in Malaysia to produce 49,000 tons of tin. The introduction of dredges and greater mechanization in gravel pump and other mines halved the total, while raising output by one-quarter. Output per man rose from 0.25 tons to 0.55 tons before the last war. Over the period 1951-5, when many dredges were rehabilitated, employment fell to 37,000, and output per man rose to 1.63 tons. Much is often made of the productivity increases in manufacturing industry, yet here we have an extractive industry in a non-industrial country achieving a more than six-fold increase in labour productivity in about 40 years.

However, in neither dredging nor gravel pumping does there seem to have been any further increase in Malaysian output per man since about 1960. Figures quoted by John Thoburn show that in gravel pumping the number of workers per 10 tons of tin fell from 21 in 1934 to 11.2 in 1950 and 7.1 in 1960, but was still 7.1 in 1972.[5] The 1978-9 figure was about 7.3. The corresponding figures for the Malaysian dredges were 5.3 workers per 10 tons in 1934, 4 in 1960, 3.9 in 1972, and 4.2 in 1978-9.

It is not surprising that output per man should have increased so much between 1934 and 1960, since horse-power consumption in gravel pump mines more than doubled, and in dredging increased by over 50 per cent. Yet, in spite of a further 50 per cent increase in power consumption by both sectors between 1960 and 1972, there was no further increase in labour productivity. According to the Tin Council report, the average labour employed per gravel pump in Malaysia was about 32 workers from 1972 to 1977, with an average output per man of 1.33 tonnes.

Since the early sixties there has been a big increase in power consumption by gravel pump mines in Thailand, but no increase in labour productivity. It is not possible from Tin Council data to determine the behaviour of labour productivity in Indonesia over a long period, but between 1971 and 1976 it appears to have risen substantially on the offshore dredges, where the main effort of the state corporation has been concentrated. Productivity has probably fallen somewhat in the onshore dredging and gravel pump sectors. The corporation's director of production and technical services is reported to have expressed dissatisfaction in 1978 with the current average output of less than one tonne per man.[6] A recent observer, however, suggests that the corporation 'regards itself as having social obligations to employ a workforce larger than necessary on strictly commercial grounds',[7] a not unreasonable policy in a developing country. The Australian industry seems to have achieved a sharp increase in output per man coincidental with a large increase in mine output.

Comparisons of labour productivity in different parts of the world industry can be made using data assembled in the Tin Council report, bearing in mind the variety of forces upon which productivity depends. Australia is clearly at the top, with an average output per man of 7 tonnes. As a very high-wage economy, Australia could maintain a competitive position only if its labour force were highly productive. At its largest mine productivity is as high as 12 tonnes per man, about five times greater than in the Cornish underground mines. The difference between the giant Renison mine in Tasmania and underground mining elsewhere is not surprising. Underground production in this mine is completely different from that of other underground mines due to the size of the seams being worked, and to the fact that the mine is worked on the level, that is, into the side of a hill with no necessity for vertical shafts.[8] The method of working is through wide tunnels which are capable of allowing extraction by 60 tonne dump trucks. At this huge mine, individual hand mining as practised in Bolivia and the UK is unknown, hence the difference in labour productivity. The large mines of the Bolivian state corporation have an output per man of 1.5 tonnes, much less than productivity in the small Bolivian dredging sector which has averaged 3-4 tonnes from 1976 to 1979. The Tin Council's figures for the onshore Malaysia, Thai and Indonesian dredges are 2.4 tonnes, 0.9 tonnes and 1.5 tonnes, and for the Thai and Indonesian offshore dredges 3 tonnes and 5 tonnes.

The less capital-intensive gravel pump mines have a lower output per man than the dredges. Malaysia and Indonesia average 1.3 tonnes,

Thailand 0.6 tonnes, probably because less power is used and output per mine is smaller. The great difference in capital-intensity of the dredging and gravel pump sectors is brought out in the Tin Council report. In the late seventies an onshore dredge in Malaysia would cost US $15,353,000, offshore dredges in Indonesia and Thailand US $35,425,000 and US $29,867,000, respectively. These figures were included in model studies for new operations, hypothetical for Malaysia and Thailand, but derived from an actual Indonesian project. In contrast, the capital cost of new gravel pump mines in Malaysia and Thailand, both hypothetical examples, was US $514,227 and US $469,000, respectively. The most expensive project was an actual Australian underground mine costing A$70 million.

It should be noted that although a dredge is much more expensive than a gravel pump, it may be expected to have a much longer working life, provided it is properly maintained and repaired. A rehabilitated dredge, as postwar experience showed, can last as long as a new one, although there will come a time when it would be cheaper to buy a new dredge than repair an old one. Moreover, working conditions may require a new dredge with a much bigger capacity, capable of digging to a greater depth. Many dredges built in the twenties and early thirties were still operating in the sixties and seventies, in many cases having been extensively rebuilt.

Competition between Dredges and Gravel Pumps

The co-existence in southeast Asia of very different methods of tin mining within the same mining districts raises the question of their relative efficiency and competitiveness. Changes have certainly occurred in the relative importance of the two main methods, dredging and gravel pumping. The share of the dredges in Malaysian output fell from over 50 per cent in the fifties and early sixties to about one-third in the mid-sixties, whereas the gravel pumps raised their share from one-third to about 55 per cent, where it remained throughout the seventies. The total output of the Malaysian dredges has been consistently below the level of the fifties for the last 15 years. Similarly in Thailand the dredges have a much smaller share of output now than they held in the fifties, although their total production, unlike that of the Malaysian dredges, has not fallen. Gravel pump and hydraulicing mines, for which only combined figures are available in Tin Council data, produced about two-thirds of Thailand's output in 1968. Since then, their share and their absolute level of production have fallen with the switch of mining activity to offshore deposits.

In both Malaysia and Thailand, Fox points out, 'the small mines showed the same vitality and power to compete successfully against the dredges'.[9] Given the opportunity, small-scale entrepreneurial activities in both countries were capable of producing large amounts of tin year after year. The situation in Indonesia has been quite different. In fact, the structure of the Indonesian industry has always been different from that of the other southeast Asian countries, even before the Dutch company was expelled by the independent Indonesian government.[10] Although Chinese miners worked in the industry, they worked directly or indirectly for the Dutch company, and not as independent operators, as was the case in Malaysia and Thailand. In postwar years there has been no scope for a similar expansion of gravel pump mining by free enterprise Chinese miners, nor by other local miners replacing the Chinese expelled from the tin islands in 1962. Moreover, the bias of the Indonesian industry under the control of the state enterprise has been switching to relatively capital-intensive offshore mining.

The relative decline of dredging in Malaysia seems to have been due to several factors. In earlier postwar years new investment may have been somewhat discouraged by the 'Emergency', but other factors have probably been much more important. Dredges require a large area to make mining a viable proposition over their long working life. With the growing exhaustion of existing large deposits, dredging companies needed exploration and mining rights to new land which were not readily granted. While gravel pump miners also had land problems, they could work successfully on small areas and on land previously worked by dredges, and they had the incentive of a relatively high price during the sixties and much of the seventies. The basic reasons for the lack of growth in the Thai dredging sector are unclear, but much less investment took place after the war than in Malaysia, and, as Thoburn notes, Thailand 'missed out almost entirely on the really large capacity "second generation" dredges built in Malaysia'.[11]

The dredge versus gravel pump issue has been examined at length by Thoburn, using several investment appraisal criteria.[12] His conclusions are generally more favourable to gravel pumps, particularly at relatively high prices, and he argues that the gravel pump sector in both Malaysia and Thailand should be actively encouraged by the government, since there is no technical obstacle to the working of many existing dredging areas by gravel pumps.

Thoburn's case for the gravel pumps includes a number of points which appear particularly relevant to a developing country. Gravel pump mining is relatively labour-intensive, an important point, it could be argued, for countries which are well endowed with labour and tend to suffer from underutilization of the labour force. Gravel pump mines are also generally locally owned and operated, giving scope for local entrepreneurs to prove that they can develop an important extractive industry which can compete on the world market with Western capitalists. There is in addition no outflow of earnings to foreign investors, and a higher percentage of inputs, other than labour directly employed, is bought locally than would be the case with a capital-intensive dredging operation, particularly when foreign-owned. Moreover, the less capital-intensive form of mining also economises the country's scarce supply of capital. As Thoburn points out, at high prices for capital (i.e. if capital in a developing country has a high opportunity cost), 'the sensitivity to the discount rate reflects the general tendency towards the greater profitability of labour-intensive techniques'.[13] But he also stresses that the problem of fluctuating commodity prices may be more serious for the gravel pumps than for the dredges. 'The failure to control fluctuations . . . can lead to greater uncertainty about future price trends and hence to capital-intensive techniques being installed in developing countries in order to reduce risks (since such techniques are viable at a very wide range of prices).'[14] Once the investment has been made, he argues, the inappropriate techniques continue to be used. This argument points to the importance of the floor price under the international tin agreements as a means of reducing the risk to the gravel pumps, an issue which will be discussed in a later chapter.

Thoburn's case seems persuasive, although it might be weakened to some extent by the huge increase in the price of energy, which is a more important cost to gravel pumps than to dredges. The growth of local participation in the share capital of dredging companies would weaken the argument about the outflow of profits. It would also have to be considered whether labour-intensive techniques were the best in the future for the opening up of much deeper, recently discovered deposits.

Costs in World Tin Mining

The broad-ranging study by the International Tin Council makes it possible to examine the cost structure of the world industry with a better statistical framework than at any time in the past.[15] From the report it is possible, admittedly with reservations which are carefully stressed by the authors, to obtain a considerable amount of information about both operating and capital costs for different production methods and different countries. It is also possible, again with reservations, to compare government claims on different parts of the industry in the form of royalties, export duties and tributes. Normal commercial costs of production are partly broken down into separate components. Inevitably, costs are simply averages for each section of the industry, which means that substantial differences between individual mines are obscured.

It is clear from the report that the highest cost sectors of the industry are the Bolivian underground mines and the Indonesian gravel pump mines, although there are likely to be significant differences between the state-owned and the larger private mines in Bolivia, with the former being the highest-cost producers. The findings on Bolivia merely confirm what has long been assumed about the cost implications of mining under the severe geological and geographical constraints of the Bolivian tin belt in the High Andes. Underground mining is shown in the report to be relatively high-cost in both the UK and Malaysia, but much less so in Australia. As expected, alluvial mining is generally low-cost, although Indonesian gravel mining seems to be the exception. As Table 2.2 shows, if royalties, export duties and tributes are excluded, the lowest cost sector seems to be Thai offshore mining, where costs as reported are well below those of other parts of the industry.

The Tin Council's data covers only about three-quarters of non-communist world production, but it shows that there is a range of just under three to one between the highest cost and lowest cost producers. A large part of the difference could be explained by geological conditions which alone would dictate different cost structures and cost levels. About one-quarter of world output, thus recorded, came from sectors with above average costs.

The report also gives information on the composition of costs for different methods of mining in different countries. It is shown, as might be expected from the nature of the technology employed, that power costs are a very substantial part of operating costs in gravel pump mining, 26.6 per cent in Malaysia, 34.3 per cent in Thailand, but only 11.7 per cent in Indonesia where diesel fuel is heavily subsidized. The difference in power costs for different methods comes out strongly in the figures for Thailand, and is particularly marked if the comparison is made with offshore dredges. The comparative figures are 13.4 per cent and 34.3 per cent. The position of Australia as a high-wage mining economy is indicated by the high share of wages and salaries, much

Table 2.2: Cost Structure of Mining Sectors as Percentage of Costs (ex. royalties, export duty and tributes), 1978

Country/method of mining	Power	Wages and salaries	Materials	Deprecia-tion	Explora-tion and develop-ment	Other
Australia						
Lode underground	4.3	31.6	31.0	13.4	4.7	15.0
Opencast	5.9	22.6	24.3	11.0	24.5	11.7
Dredge	8.9	50.7	23.0	7.8	1.6	8.0
Bolivia						
Lode underground	3.2	32.8	15.5	2.4	4.2	41.9
Dredge	4.7	19.9	17.6	6.8	2.9	48.1
Indonesia[a]						
Dredge offshore	10.3	22.3	18.7	3.7	3.4	41.6
Dredge onshore	11.7	27.6	18.2	4.4	2.9	35.2
Gravel pump	11.7	30.6	18.5	4.5	4.1	30.6
Malaysia						
Dredge	20.4	23.3	21.6	10.5	0.4	23.8
Gravel pump	26.6	30.6	17.4	8.8	1.6	15.0
Open cast	7.1	51.4	5.2	7.4	—	28.9
Underground	12.9	51.0	21.5	5.6	0.3	8.7
Thailand						
Dredge offshore	13.4	23.0	22.2	10.5	—	30.9
Dredge onshore	26.4	31.5	22.0	6.9	0.1	12.1
Gravel pump	34.3	26.8	18.2	6.6	2.2	11.9
UK						
Underground	8.1	53.1	16.3	5.4	9.7	7.4

Note: a. Indonesian costs are helped by a very substantial subsidy on diesel oil, which the mining industry buys at well below world prices.

Source: ITC, *Tin Production and Investment*, p. 137.

higher than in the three southeast Asian countries. It is interesting that the area with the highest share of wages and salaries is Cornwall.

The report emphasizes that the breakdown of costs has to be treated cautiously, since much depends on the relative prices paid in each country for factor inputs and on different accounting procedures, notably on different ways of treating depreciation. Further, in both Indonesia and Malaysia, a large part of costs has been put under the heading of 'other charges', which do not include power, materials or depreciation, each of which is given separately. Depreciation in Bolivia lode mining is given a very low figure, well below the percentages for both Cornish and Australian underground mining.

Economic Rent, Taxation and the Government Share

The addition of royalties, export duties and tributes to the production cost figures makes a considerable difference to the ranking of producers. Bolivian underground mining becomes the highest sector by a small margin over the Indonesian gravel pumps. The most obvious change is that UK and Australian mining costs are now brought well short even of costs in Malaysian gravel pump mines, and become fairly comparable with costs in southeast Asia in general. This is simply the result of the much smaller claims in the above forms by government in the two developed countries for a share in the economic rent from tin mining (see Table 2.3).

In all the developing countries listed, and no doubt also in Nigeria and Zaire, two other members of the Tin Council, the government skims off a large part of tin revenue. Figures supplied to the Tin Council show that net revenue (price less cost) is halved in Malaysian dredging, and cut by about 60 per cent in Thai dredging. Net revenue of the Malaysian gravel pumps is reduced by over two-thirds, and of the Thai gravel pumps also by about 60 per cent. The net revenue of the high-cost underground mines in Bolivia is converted into a small deficit. The one large Malaysian underground mine, with a pre-tax net revenue similar to that of the UK mines, apparently suffers an 80 per cent cut. Only a moderate reduction in net revenue is experienced by the nationalized Indonesian dredges.

As the Tin Council report points out, the difference between costs, including government claims, and the selling price has to be used in a number of ways: amortizing and servicing loans, exploring off-site for new workable deposits, paying corporate, income, or profits taxes, maintaining and increasing the capital stock, and paying dividends. Profitability is clearly much affected by differences in government claims. Variations in the government share as a percentage of gross profit shown in the estimates highlight the differences in tax systems. It appears that in Australia royalties and export duty are a very small percentage of the price of tin, only 2 per cent compared with up to 29 per cent in Malaysia, 30 per cent in Thailand, 35 per cent in Bolivia, and 16 per cent in Nigeria. Expressed in terms of Malaysian dollars per pikul of contained tin (i.e. the metal content of ore), fiscal charges at a price of M$2,000 a pikul ranged from M$40 in Australia to M$698 in Bolivia. (A pikul equals 133.33 lb or 60.479 kg.)

The magnitude and incidence of these fiscal charges, plus income taxes, have long been a source of controversy between the developing producing countries and some leading consuming countries. They have

Table 2.3: Production Costs and Economic Rent[a]

	Country	Method of mining	M$/pikul			
				Cost[b]	Price less	Cost[c]
			(−R)	(+ R)	(−R)	(+ R)
1.	Thailand	Dredges offshore	545	1,060	1,198	683
2.	Malaysia	Dredges onshore	595	1,094	1,148	549
3.	Malaysia	Opencast	614	1,117	1,129	626
4.	Thailand	Dredges onshore	644	1,157	1,099	586
5.	Indonesia	Dredges offshore	664	834	1,079	909
6.	Thailand	Gravel pumps	793	1,320	950	423
7.	Australia	Lode underground	850	877	893	866
8.	Bolivia	Dredges onshore	953	1,456	790	287
9.	Indonesia	Dredges onshore	971	1,142	772	601
10.	Malaysia	Gravel pumps	988	1,530	755	213
11.	Malaysia	Underground	1,152	1,658	591	85
12.	Australia	Alluvial dredges	1,156	1,197	587	546
13.	UK	Underground	1,175	1,225	568	518
14.	Bolivia	Lode underground	1,358	1,772	385	−29
15.	Australia	Opencast, lode, alluvial	1,378	1,420	365	323
16.	Indonesia	Gravel pumps	1,564	1,734	179	9

Notes: a. Economic rent (+ R) includes royalties, export duties and tributes.
b. Weighted average 1978. Costs within each group differ widely from the average.
c. Average price, Penang ex-smelter for 1978 = M$1,743/pikul.

Source: ITC, *Tin Production and Investment*, p. 127.

also been a long-standing grievance among miners in the producing countries themselves. Where the enterprise is state-owned, it is a question of the division of the revenue between the enterprise and the government, the latter claiming to represent the interests of the country as a whole, and taking into account the role of the industry in the national economy. The effect on the industry depends on a number of factors. There is no doubt that in tin mining, as in mining generally, there are great differences in the excess of price over cost between different areas and mines. These differences reflect factors such as the richness and accessibility of deposits, the efficiency of the labour force and management, and the percentage of metal realized from the ore. Some mines, therefore, are highly profitable, more or less irrespective of the government's charges; others are just on the margin of profitability at the level of taxation.

The closer the market price to the costs of the marginal producers, the larger the surplus over costs that accrues to the intra-marginal producers. In some cases, the surplus may be very large. Over a substantial part of the industry in southeast Asia this surplus or economic rent could be over half of gross revenue.

Conceptually, it is possible to distinguish several types of economic rent.[16] One type, associated with the name of the well-known nineteenth-century political economist David Ricardo, can be described as a Ricardian-type differential rent, which arises from differences in the location of mines or in the quality of their ores. Some Malaysian and Thai alluvial deposits give miners very low costs in relation to those of much of the world industry. Thus Ricardian rent plays a big part in the earnings of these mines. Although their deposits have certainly been declining in quality, this has probably not affected their relative position on a world scale. Hence this type of rent may be long-term in nature.

A short-term or windfall rent occurs in mining at intervals, chiefly as the consequence of sudden increases in demand, when the short-term elasticity or supply is low. This has happened in the world tin market several times in the last couple of decades.

A third type of economic rent can be regarded as a monopolistic surplus which may arise in two ways: either through the operation of a cartel such as existed in the tin market in the thirties, or through some aspect of the structure of the market.

A fourth type might best be described as a quasi-rent. This may arise from the ownership of capital in an industry, the supply of which cannot be quickly increased in the short run, or from managerial or technical expertise, which may confer a long-run advantage on the possessor. The uniqueness of this type of rent lies in the mobility which would enable the possessor to escape if conditions became too unfavourable.

A fifth type of rent is the scarcity rent which is associated with the exhaustibility of a non-renewable resource in the case of a mining industry. Its relevance to tin, or indeed any mineral, fuel or non-fuel, depends on the reserves position or on expectations of an adverse change in the reserve position.

Since, by definition, economic rent is a surplus over cost, including under cost an adequate return on capital, with allowance for risk, it could in principle be taxed away by the government without affecting either the continuation of enterprises or investment in new enterprises. Attempts, however, to tap quasi-rents would involve discrimination against factors of production which could escape. Governments of developing countries, or indeed of developed countries, may well underestimate this possibility. They might reasonably assume that the objective of their mineral tax policy should be to capture all scarcity, differential, monopolistic and windfall rents, while leaving the investor in the position to make an adequate return. There is, unfortunately, the problem that government and business views on what constitutes an adequate return, including an allowance for risk – particularly important in mining – may not coincide. If they are too far apart, the volume of mining activity will be reduced. If quasi-rents are involved, there will certainly be a loss of managerial and technical expertise.

It can be argued that until recent times the governments of developing countries often obtained a smaller share of economic rents from the exploitation of their mineral resources than their economies required. In some cases this might have been due to the political relationship between governments in the colonial age. It might also have reflected the general environment and economic thinking which were more tolerant of free enterprise in mining.

Over the last two or three decades, major changes have occurred in the attitude of governments and their electorates to the exploitation of mineral resources within their control. The most dramatic have been the revolutionary change in the treatment of petroleum revenues by the OPEC. However, changes have by no means been confined to developing countries, as the experience of mining companies in Canada has shown. It is in fact a widespread phenomenon, affecting all mineral-producing countries. Nevertheless, in a recent study of mineral rent from minerals, Helen Hughes and Shamsher Singh came to the conclusion that 'only in tin, and to a lesser extent, petroleum, have the exporting countries been able to capture a large share of the total rent'. As a measure of rent, the study took the sum of the royalties, taxes and duties levied, or the difference between f.o.b. prices and production costs. In order to illustrate the division of economic rent arising from tin, the writers used trade between Malaysia and the US, between 1971 and 1976. Their calculations implied that total economic rent in 1976 was US$2,342 per tonne, of which the Malaysian share was estimated to be 84.5 per cent and the US share 15.5 per cent. Although there were fluctuations in shares, according to their data Malaysia appeared to do well in every year.[17]

It is now a high priority of governments in developing countries to push for general economic growth, which inevitably means that far greater emphasis is placed on the contribution which the mining industry can make to diversification of the national economy. It has also meant that governments have become much more restrictive in their policies towards the role of foreign-based companies in tin mining and in mining generally. At the limit they have simply nationalized foreign enterprises.

Another major change has been the awareness that mining inevitably depletes a fixed asset (however uncertain its ultimate size), which is the source of economic rent. Governments, therefore, aim to maximize their revenue from the fixed asset, whether or not it is exploited by a foreign company. They also insist on a large tax income from state enterprises partly because there is a tendency for wages and consumption to evolve as if economically exploitable mineral resources would last indefinitely. Unless this tendency is checked, serious difficulties are possible when the apex of the mineral-rent cycle is reached.

How damaging a tax system may be to the mining industry is a controversial issue. Different interests naturally take different views. The Tin Council report treads warily in the minefield, as befits an organization endeavouring to reconcile a variety of producer and consumer interests. It points out that 'there may be reasons for the preference given to governments to industry other than tin mining'.[18] More specifically, it agrees that 'the opportunity cost of investment may favour some other uses of land, such as rubber production, or a government may have a policy of diversification to avoid over-reliance on a small group of commodities, or it may restrict mining for environmental reasons'.[19]

The report then gives reasons why governments should exercise some restraint in the limitations it imposes on tin mining and in the choice of a tax system which may discriminate against mining companies, whether foreign or national. These are, firstly, 'the safe-guarding of tin deposits against premature abandonment because the after-tax rate of return to mines has become too low', and secondly, 'the secondary economic benefits accruing to the national economy as a result of mining activities, for example, foreign exchange earnings, the employment of labour, purchase of materials and equipment from local industries and the use, generally, of local resources such as power and fuel'.[20]

The mining industry in developing countries tends to have a bad image, which often obscures the merits of its case for fair treatment. Politically, it often lacks the support of pressure groups such as those behind industrial activities in urban areas. The long-term adverse effects of a bad tax system are often underestimated. Moreover, as an export industry par excellence, mining qualifies as an easy and remunerative target for an inefficient tax system. An export duty, for example, is a simple tax to collect and difficult to avoid, except by smuggling, which has been a significant side-line in southeast Asia during the seventies, but for the great bulk of the industry the duty is inescapable and adds to production costs.[21]

The various issues and complexities of mining taxation in the world mining industry cannot be examined in detail, especially as the tax systems in the tin producing countries differ considerably. Whether tin mining suffers disproportionately is uncertain. The question of discrimination against tin was considered by the authors of the Tin Council report. Although there was some evidence of discrimination, it was not conclusive. In Bolivia it appeared that in general 'tin, antimony and wolfram were less favourably treated than other minerals in respect of the combined payment of export tax and royalty'.[22] There seemed to be a slight disadvantage to tin in the Indonesian system, but no meaningful comparison could be made in Malaysia, although there were some aspects of the tax system which seemed open to criticism.

Information available to the Tin Council showed a big increase in fiscal charges on tin mining in three developing countries between 1971 and 1978. The report compared estimated cost increase in six producing countries between these years, both including and excluding fiscal charges. While production costs had increased substantially in both the UK and Australia, as in the four developing countries, there was a very marked difference in the additional fiscal charges between the two groups of countries. Not much change had occurred in the two developed countries, whereas the cost index, including fiscal charges, rose sharply in Bolivia, Malaysia and Thailand. It was clear that these three countries, in the conditions of the seventies, found it necessary to exact a much bigger contribution by way of indirect taxes from tin mining. The share of royalties and export duties in total costs in Malaysia and Thailand rose sharply between 1971 and 1978. There was also some increase in Indonesia.

Whether these taxes were pushed too far in the seventies is debatable. So far as Malaysia is concerned, Thoburn takes the view that, as a deterrent to investment, taxation has been less of a problem than the government's restrictions on the availability of land for mining, a point discussed in a latter section of this chapter.[23] According to the authors of the Tin Council report:

> Without in any way diminishing the right of countries to sovereignty over their natural resources and their rights enshrined in UN Resolution 3203 to make the development and exploitation of those resources serve their national interests, there does appear to be room for discussion on how these national interests are best served, not only in respect of the quantum of rent charged . . . but also in respect

of the manner in which the rent is charged, i.e. the balance between indirect taxes in the form of royalty or export charges, which can affect the cost of production, and direct taxes in the form of income tax or corporation tax, which affects the profitability of production.[24]

The issue is clearly a highly sensitive one for both domestic and external relations. A particular problem arises with royalties, which are a popular charge with governments of developing countries. Royalties on each tonne produced are not related to costs and have the important disadvantage that they affect the cut-off grade of commercially profitable deposits and hence the marginal level of reserves. To put the matter bluntly, governments can convert 'ore into rock' by their tax policy.

The problems of mining taxation in a number of developing countries, including several tin producers, have been examined recently by an American study group.[25] The dilemma facing a revenue-hungry government is well illustrated by Bolivia, which is particularly dependent on tin mining for both tax revenue and export earnings. The tax burden on the Bolivian industry, including the nationalized COMIBOL, is believed to be among the world's highest. The Gillis inquiry, pointing to a serious lack of investment, attributed it to the disincentive effects of both the structure and the level of taxation. Four main shortcomings of the system are identified: firstly, the system as a whole is too burdensome for the industry in bad times, yet paradoxically does not benefit the government sufficiently when times are good; secondly, perverse incentives for exploitation and concentration decisions lead to socially wasteful 'high grading' of ore deposits and excessive losses in the concentration process; thirdly, there are deterrents to investment in risky activities; fourthly, equal treatment of enterprises which are vastly different in cost structure tends to undertax significantly large mechanized operations and overtax more primitive operations. The Gillis study estimated that Bolivian export and/or output taxes as a percentage of export value in 1974, based on an LME price of US$4.00 a lb., were 31.72 per cent, compared with 22.6 per cent in Thailand, 16.88 per cent in Malaysia and 10 per cent in Indonesia.

Paradoxically, although these taxes appeared to be a heavy burden on Bolivian mining, Bolivian land rents on mining concessions were found to be extremely low. The land rent is paid by all holders of exploration or exploitation rights in Bolivia. As a percentage of mining revenue collected by the government, the proceeds of the land rents are less than one-tenth.

The significance of low land rents is that they encourage claim holders to keep back promising mineralized areas from exploration and possibly exploitation. In fact, according to the Gillis study, land rents are so low that they positively encourage speculators to indulge in mine-hunting. Gillis comments: 'Since annual land rent levels for a medium-sized exploration claim (10,000 hectares) amount to only US$600 and since it is a simple task for one person to control as many as ten medium-sized claims, the holding of large tracts of idle claims is not an unduly expensive proposition'.[26] This potential mining land is kept out of use. Whether this is entirely wrong from the national point of view is debatable. It depends on the relationship between present and future discounted tax revenues from the minerals. However, there is also another large area of potential mining land out of circulation. The Bolivian government holds a large part of the mineral belt in the so-called fiscal reserves, that is, land from which future mineral production may be expected to contribute to government revenue. Together, the fiscal reserves and the land affected by the land rents, mean that much of the most promising areas has been unavailable for exploration, according to the Gillis study.

There has been extensive debate in Bolivia on the defects and possible reform of the system of mining taxation. The basic system of assessment is the 'costo presunto', which is the officially established cost of production. This is an assumed cost, used by the government to determine a company's profit, arrived at by comparing the 'costo presunto' and the world tin price. If the 'costo presunto' is too low, the tax is relatively high. There are various complications in the system, leading to arguments about whether the end result is fair or unfair taxation and damaging to production.

Gillis points to another weakness in the Bolivian mining tax system. Discussing whether the system discriminated adversely between mines, he argued that much would depend on the variance in costs among mines of the same category. A tentative survey of COMIBOL and medium-sized mines suggested a considerable variance in costs, much of which was due to differences in exploration and mine development. Under the existing tax system, according to Gillis, the mine which had spent large sums on these activities paid the same taxes as the mine that spent none.

Whether miners will actually explore more if they pay less taxes because of expenditures made on exploration is a moot question. However, private miners indicated great eagerness in securing exploration money from abroad. That money is supposedly more forthcoming because the foreign investor can deduct such expenditures in calculating his income tax at home.[27]

Virtually stagnant production in Bolivia suggests that more could be done to help the mining industry, and thereby the present low-income citizens of one of the poorest countries. Moreover, as Gillis points out, the mining community has the technical abilities to make a success of the industry in its difficult physical environment: 'most managers [of COMIBOL] have a strong background in engineering, and both anecdotal evidence and observation indicate that technical competence has been high from almost the birth of the enterprise, especially in relation to other public and most private sector firms in Bolivia'.[28] On the basis of this judgement it is certainly disappointing that the state industry has not been allowed the resources to give a better performance in the years since the tin baron mines were nationalized.

Long-run Trends in Production Costs

Technological innovations and their effects on production methods, and changes in the relative importance of different production methods and producing areas have affected trends in costs and prices over the years. The movement of Chinese miners into Malaya after the middle of the nineteenth century led to improvements over the methods hitherto used by Malay miners. From the seventies, the expansion of alluvial mining in southeast Asia brought about a drastic decline in the long dominant Cornish industry. Eventually it also halted the growth of the young Australian industry, which then went into a decline from which it did not recover until the 1960s. After the First World War the large-scale introduction of dredges and increased mechanization of other forms of mining strengthened the position of the alluvial section. That these developments in the twenties did not lead to much lower prices in the thirties, apart from the worst period of the depression, was due to market intervention by the tin cartel, which checked the effects of lower costs on prices.

Before the expansion of alluvial mining in southeast Asia, there were clear signs that the growth of demand was putting pressure on the price of tin. From 1850–1 to the mid-fifties the price rose by over 40 per cent, and remained from 40–65 per cent above the earlier level until

1864. There was a temporary intermission from these relatively high prices in the later sixties until 1869, when the price suddenly rose by about 20 per cent. From 1869 to 1873 the price was generally high, reaching a peak of £150 a tonne in 1872, a level not reached again until 1906.

By the late seventies the production of low-cost alluvial tin led to a drastic fall in price, which exceeded £100 a tonne in only three years between 1875 and 1898. The period 1894 to 1898 saw the lowest prices in the second half of the century. This was the period in which the fastest growth occurred in southeast Asia, accompanied by the sharp fall in output from the higher-cost Cornish and Australian mines. Between the eighties and nineties production doubled in British Malaya and the Dutch East Indies. A 70 per cent increase in price in 1898–9 followed this period of expansion, after which there was no return to the price levels of the last quarter of the nineteenth century. Relatively high prices persisted throughout the first decade of the present century, with even higher prices for a few years before the First World War.

At this time, important cost-reducing innovations which were to revolutionize the industry were still in their infancy. The first dredges were just beginning to make an impact on alluvial mining in southeast Asia. Dredging accounted for only about 5 per cent of Malayan output in 1915 and 13 per cent in 1920. Although Western mining techniques were also becoming more important in hydraulicing, most Malayan output, then about 45 per cent of world production, still came from Chinese mines, which were in the process of becoming more mechanized.

After the postwar slump there was a period of very high prices which encouraged a major investment boom, the result of which was eventually a substantial fall in costs over a large part of the industry. It has been estimated that operating costs of the mechanized alluvial mines in southeast Asia and Nigeria were nearly halved in the late twenties.[29] According to the same authority, even in the relatively high-cost Bolivian mines, where grades of ore were already falling from the high levels of the late nineteenth century and early years of this century, there was an appreciable fall in production costs, excluding depreciation and taxes.

Since the Second World War there has been no comparable fall in costs. As a previous section has shown, there has been no increase in average output per man in Malaysia or Thailand since the sixties. It is unlikely that Bolivia and Indonesia have done more than recover some of the ground lost after the difficulties associated with the changes in ownership and control during the fifties. With no major technological changes for several decades, production costs in the industry as a whole have probably risen, which would be compatible with the long-run upward trend in the price of tin. The Tin Council report confirms this trend, which Smith and Schink have also identified over the 20-year period from 1956 to 1975. The position was not clear in the earlier postwar years because of the shortages of other metals and violent price fluctuations. Since the fifties, however, there has been a noticeable shift in relative prices.

For many years the producing members of the International Tin Council have complained that their rising costs have not been sufficiently recognized in the price limits specified for the buffer stock range. One of the chief elements in the producers' case is the fall in the grade of ground ore with which they had to work. They have repeatedly argued that it has become impossible to offset a more or less persistent fall in the grade by improvements in production techniques, which implies that the real cost of tin mining must rise.

Grades of Ore and Production Costs

That a falling grade of ore should mean higher costs, other things remaining equal, seems a reasonable proposition. Various aspects of the problem have been examined closely by the recent Tin Council report, which points out, however, that 'this is an area in which it is difficult to make an assessment with any precision'.[30]

There is nothing unusual about a fall in grade of ore, either in tin or any other non-ferrous metal. Much evidence is available of falling grades of ore in copper, lead and zinc. In many important mining areas grades are well below those common earlier this century and in the first quarter of the present century. The well-known study by Orris C. Herfindahl of long-run trends of copper costs and prices showed how the grade of copper ore mined in the US, the world's largest producer, had fallen dramatically from 1879 until the late fifties.[31]

Ore grades in tin mining have generally been lower than those of other non-ferrous metal mining industries. Tin miners, as pointed out by Fox, work alluvial ground with a metal content of around 0.017 per cent tin, and lode deposits with 1 per cent. Zinc ores, on the other hand, are generally in the 4-16 per cent range, lead ores 3-12 per cent, copper ores 0.4-0.6 per cent. According to Fox, 'substantially higher figures for quality of ground in other metals, except gold, are a partial explanation for the relatively high price of tin'.[32]

The general impression of tin ore grades is that they have been falling for many years. A large amount of information on grades throughout the world tin industry was submitted to a Tin Council Working Party in 1962-3.[33] Big reductions were reported in Bolivia since the days when phenomenally rich veins were exploited by private companies. At the largest Bolivian mine, which had produced more tin than any other mine, the average grade of ore had fallen dramatically from the exceptionally high levels of 12-15 per cent tin to less than 3 per cent by the late thirties and to only 0.53 per cent in 1963. The average grade of ore treated at the COMIBOL mills in 1964 was 0.82 per cent compared with 3 per cent in 1938. According to the latest Tin Council report, the average grade of Bolivian ore is now about 0.72 per cent.

In a study of the Bolivian industry, David Fox has pointed out that a different method of underground mining, adopted to cope with falling grades, would cause a reduction in the average grade actually mined, since the definition of a workable grade would fall. He notes, however, that a great change has occurred over the years in the accessible grades of ore in Bolivian mines, with the lower grades being worked appearing at greater depths and presenting more difficult mining conditions.[34]

The Tin Council report states that the grade of ore in the largest Australian mine, responsible for about half the country's output, was nearly twice the Bolivian average, and nearly three times that of the largest Bolivian mine in the early sixties. The average UK grade was also higher than the Bolivian average in 1978.

In the alluvial mining countries of southeast Asia, the grade of ground in Malaysia was reported in 1963 to have fallen considerably since the thirties. Further reductions seem to have occurred in the seventies, according to the latest Tin Council report. No information of the position in Thailand and Indonesia was available to the earlier Working Party, but the latest report suggests that grades have not been falling in Thailand in recent years. For Indonesia there was a good chance that offshore deposits, which were of growing importance, would have high grades. Unfortunately there was no information on the likely grades in the large prospective mining area of Kuala Langat in Malaysia, the largest discovery for many years.

The overall position on ore grades is not easily summed up, since there are differences throughout the industry and important gaps in the data. However, there can be no doubt that grades are much lower in Bolivian and Malaysian mines than they were before the Second World War. This is probably true also of the African producers and

Indonesian onshore mines. In Malaysia many gravel pump mines are working ground which has already been mined two or three times. At the large Bolivian mines, mountains of tailings contain ore grades well above those now being mined underground. Eventually technical progress will make it possible to recover much of the tin locked up in these tailings.

It is possible that a higher rate of exploration over a number of years would have had a favourable effect on the ore grades now available for mining. In the copper industry, where much more is spent by the large mining companies than is the case in the more diversified tin industry, there are indications that grades are not falling at anything like the rate of the last few decades. There is evidence that deposits which have been opened up in recent years do not differ much from older deposits which are already being mined. Nevertheless, it would need an extensive addition of new deposits to make much impression on the average grade, which naturally reflects the falling grades of the last few decades.

Lower grades involve increasing inputs of capital equipment, materials, energy and labour, unless there are compensatory improvements in technology. This is a crucial qualification. The quality of mineral deposits is only one determinant of mining costs. If this were not so, Herfindahl's findings on copper costs and prices in the US would have been very different. Herfindahl showed that in spite of a long-run fall in the quality of American copper ores, there had been no rise in the real price of copper between 1880 and the late fifties. Copper had not become more expensive in relation to the prices of a broad range of commodities included in the US wholesale price index. According to Herfindahl, 'after the elimination of the abnormal years before World War One, copper prices seem to have been comparatively stable over a long period. There are sizeable gaps in the record of normal years after World War One, but even here, the remaining years yield no suggestion of upward or downward drift over the period as a whole'.[35]

It follows that there must have been offsetting cost reductions in American copper mining. The source of these cost reductions was a striking advance in mining techniques. The mining industry adapted itself to working on a much larger scale with lower ore grades. It was essentially major changes in mining technology which were responsible for reducing the real cost of copper from the 1880s to the 1930s. Similar improvements came into effect in the mining of lead, zinc, bauxite and iron ore. Since low-grade deposits were much larger than high-grade deposits, it was possible to achieve much greater throughputs with modern technology. In postwar years costs have been kept down by using massive excavator and transport equipment. Long-distance transport costs for iron ore, for example, have been cut by bulk ore carriers.

Consider now the position in tin mining. Before the First World War the problem arose of mining alluvial deposits at a greater depth and with a larger throughput of material. The introduction of the dredge solved this problem. The average monthly throughput of the early dredges, around 57,000 cubic yards, has been far exceeded, with modern large-capacity dredges in the seventies handling 15 times as much material, and new dredges planned with throughputs of over one million cubic yards. These developments in the dredging sector were the necessary response to the exhaustion of the richer and more accessible deposits. Lower grade deposits lay at greater depths and were overlaid by a large amount of barren overburden, which had to be removed before payable ground was reached.

Technical improvements have also been made in gravel pump mining to cope with lower grades of ground. It is an indication of the progress made by gravel pump miners that the grades worked in the seventies could be as little as one-tenth of those that were considered average at the beginning of the century. In mining much lower grades it has been necessary to use a much larger amount of energy per unit of output. Horse power consumption for every ten tonnes of tin rose more than three-fold between 1938 and 1972 in gravel pumping, and nearly doubled in dredging.[36]

The expansion of Australian underground and opencast mining since the mid-sixties has been possible only by the treatment of low-grade deposits on a large scale. New mines were opened up by large companies using the latest mining techniques giving high labour productivity. The general mining environment was comparatively favourable to investment, in contrast to the position in the other major area of hard-rock mining, Bolivia, where the industry has been particularly vulnerable to falling grades.

Although there is no doubt that much has been done in parts of the world industry to cope with lower grades, it seems that tin mining has been less able than other non-ferrous metal mining to escape the effects of lower grades on costs. To some extent this is probably due to the nature of tin deposits. The rest of this chapter suggests that other influences have played a part. While the size of dredges is still increasing and more mechanization has been adopted throughout the industry, the average output per mine has not risen significantly for some time, as it has done with other metals. However, there are signs of change, with much larger gravel pump mines in Malaysia, according to the Tin Council report. Further, the scale of operations is certainly rising offshore in Indonesia, where the latest additions to the offshore dredging fleet can each increase Indonesian output by at least 5 per cent.

There are other factors which could influence the trend of costs, but it is not axiomatic that tin mining should suffer more than other mining from them. Complaints have come from developing countries about the prices of imported inputs from manufacturing countries. A wide range of spares, materials and explosives, particularly important for the underground Bolivian mines, have become much more expensive as a result of the persistent inflation in the industrial supplying countries. According to Gillis,

> the cost of explosives alone has accounted for about 30 per cent of the per unit value of tin leaving the Catavi concentration plant in August, 1975, following the marked increase in ammonium nitrate prices and the 1975 decline in world tin prices. This leaves only 70 per cent of value to cover costs of other material inputs, labour and energy costs, taxes, depreciation, transport, smelting and other costs.[37]

Among the developing producing countries, Malaysia is the only one with a substantial local industry which can supply equipment for the mines, but it still depends on imports to a large extent. Figures compiled by the Tin Council, covering six countries, show large cost increases, excluding royalties and export duties, ranging from 180 per cent to 49 per cent, for the 1971-8 period. Analysing the data for the period, the report comments: 'There has been no consistency or pattern in the changes which have taken place over the various sectors which make up the cost structure of the industry'.[38] Even so, these price increases recorded by the Tin Council do not explain the long-term trend. The same conclusion applies to the impact of higher energy prices. It might be expected that in the seventies the rising cost of energy would hit tin mining, or at least a substantial part of the industry, since it is probably more energy-intensive than mining industries producing other metals, except aluminium, which is produced from bauxite with a prodigious use of electricity. Nevertheless, looking at the energy position before the OPEC exploited its monopoly power, tin mining should have benefited from the years of cheap energy.

Exploration

Faced with declining ore grades, producers of other minerals have ranged widely throughout the non-communist world in search of new, preferably higher-grade, deposits. Spending on exploration has been a substantial part of mining company budgets. The results of these efforts by large American and European multinational mining companies have been a major transformation of the world mining industry since the Second World War. New producing countries have come into prominence. Mining has become more capital-intensive and shipping costs have been reduced.

In tin mining, spending on exploration seems to have been patchy for many years. During the thirties there was considerable spare capacity as a result of the previous investment boom, which inevitably discouraged new investment. The earlier postwar years were a period of rehabilitation in southeast Asia, remarkably successful in expanding throughout but depending heavily on US strategic stockpile buying in the late forties and early fifties. Indonesia's political and economic difficulties after the nationalization of the Dutch-operated mines meant years of uncertainty and decline, with the state-owned mining corporation lacking the resources to invest much in exploration. It is difficult to assess the extent of exploration in Malaysia since it is scattered over a number of agencies. An indication of the problem is the fact, noted by the Tin Council report, that Malaysia has never made a comprehensive survey of its tin reserves. It is also striking that the money allocated by the government for exploration and research in the mining sector has usually been less than might be expected from the importance of the Malaysian industry both nationally and internationally.

In all the tin producing developing countries the agencies involved in the mining industry seem to be under-financed and under-staffed by comparison with advanced mineral producing countries like Canada and Australia. This is not a peculiarity of the tin industry. It is a general characteristic of all developing countries with a mining sector, which has to compete with other, generally more vocal and politically influential claims on the national resources.

The Bolivian industry seems to have suffered particularly badly from under-investment in exploration over several decades. The Gillis inquiry found that 'data on total realized investment in exploration, mine development, exploitation and concentration are scanty at best',[39] but there seemed no reason to doubt the lack of investment. A Bolivian estimate of US$10 million spent on exploration over ten years compared unfavourably with exploration outlays on *single* claims by a number of individual companies in Indonesia over the period 1968-75. Gillis found that the medium mining companies, the most dynamic mining sector, had not invested much new capital since 1968-9. As for the state corporation, COMIBOL, 'investment in exploration . . . has largely followed an erratic pattern, characterized by long delays and unforeseen difficulties in implementation'.[40] Gillis gives, as an example, one of the most successful discoveries at one of the largest mines which came up with 'a dozen new veins with average widths of one metre (wide by Bolivian standards) and of tin content of between 1.5 and 3 per cent compared with present grades in Catavi of less than 1.50 per cent'.[41] This project, initially approved in 1963 to be completed in two years, took 12 years to put into effect. Gillis concludes gloomily: 'the mining sector has proved to be resistant even to the surge in mineral prices in 1973 and 1974'.[42] This picture of low investment has to be seen against a background of inadequate investment for years before the 1952 revolution, since the former owners of the large mines, with the threat of nationalization hanging over them, were more interested in 'gutting their mines before nationalization'.[43]

In the three smaller producing countries, Brazil, Australia and the UK, there has been much more exploration activity, with favourable effects on output. Total output from these sources, however, is only about one-third of the output from Malaysia, and barely made up for the decline in Nigeria and Zaire during the seventies.

Access to Mining Land

Prospecting for new tin deposits, and tin mining generally in Malaysia, have been handicapped for many years by difficulties in obtaining access to land. There have been frequent complaints from miners that it has been extremely difficult to get an exploration licence. As far back as 1952, the Paley Commission, referring to obstacles in pre-independence Malaya to the extension of tin mining beyond existing mineral-rich areas, reported that future tin production depended to a considerable extent on solving this impasse.

The fact that Malaysia is a federal state is a source of problems for the mining industry. Difficulties always arise in a federal state over the division of economic power between the federal and the state governments. The Malaysian state governments have the responsibility for issuing the necessary prospecting licences, but the federal government has obtained most of the tax revenue from subsequent mining operations. The state governments have had a long-standing grievance that their share of the mineral wealth arising within their boundaries has been disproportionately small. Thus they had little financial incentive to encourage mining, especially if there was a possibility at some future date of allocating mineral-bearing land for housing or for industry, the latter being capable of bringing them a larger revenue.

Particularly important is the fact that state governments might also incur costs in resettling smallholders or other people in an area allotted to mining. 'Resettlement is an explosive issue and one where the state governments are anxious to avoid trouble at all costs, especially when it is a question of allowing a foreign-owned company to deplete the state's mineral resources'.[44] The situation is complicated by the fact that it is usually the Malay population who are affected by the alienation of land for tin mining. One of the most jealously-guarded rights of the states is their complete authority over the development of their land, an authority which is enshrined in the Malaysian constitution. This means that the states have legal title to all unclaimed land, as well as the freehold of most privately-acquired land. From this authority vested in the state governments it follows that it is the responsibility of the governments to vet all applications which involve a proposed change of use, whether the land is to be developed for housing, mining or industry. As one critic has emphasized, 'The industry's problems over the past ten years have been political rather than geological. The mining companies have been caught between the aspirations of the state governments and the wider demands of the federal government in Kuala Lumpur'.[45]

Difficulties in obtaining titles to mining land have affected both dredging and gravel pump companies. According to one dredging company:

> Delay in acquiring mining titles is a major problem. We have applied for mining titles over 332 of the 421 acres of agricultural land the company holds under lease, but approval has so far been obtained over two small areas totalling 11 acres. The most recently acquired mining certificate took three years from the date of application to the actual issue of the title enabling the dredge to work the land.[46]

That delays should occur, follows inevitably from the number of hands through which applications go before a company receives authorization. The Tin Council report, quoting a paper delivered at a national mining seminar, states that '17 persons or bodies in up to 10 State

Departments must give reports/recommendations before a licence/permit can be issued'.[47]

Prospecting on state land in Malaysia can be carried out under one of two systems. The prospector may acquire a prospecting licence which gives the holder the right to prospect and also the prior right to select any part of the land for mining, with the automatic entitlement to a mining lease under such terms and conditions as may be imposed by the state Ruler in Council. Alternatively, the prospector may obtain only a prospecting permit, which gives the right to prospect, but does not convey any right to the eventual issue of a mining lease. It has been stated that prospecting licences are uncommon, that permits are often issued to people who have no intention of prospecting themselves, and that 'even when licences are issued, there is no assurance that the terms and conditions attached to the eventual issue of a mining lease would be such as to encourage prospecting and eventual exploration'.[48]

Thailand appears to put fewer obstacles in the way of would-be prospectors, although there are, of course, limitations on foreign companies. The Tin Council report states that 'there is generally felt to be no obstacle to the expansion of mining, except where possible tin values are likely to provide a lower economic rent than would be available from other land usages'.[49] The location of Bolivia's mineralized area implies that there is no competition from other possible land uses. There is a problem in Nigeria, however, where tin mining has been carried on in a comparatively small area, which has considerable competition for available land from the expansion of two towns.

No discussion of the availability for tin mining would be complete without a reference to the impact of environmental consideration. This is a comparatively new phenomenon in the mining world, and one which can add substantially to production costs although the effects are likely to vary between minerals and countries.

The nature of alluvial mining in Malaysia and the location of a high proportion of the country's deposits within or near populated areas, have led to tighter regulations over environmental damage than were customary before the war. Public authorities have also been influenced by the opportunity cost of mining land, which may have an agricultural use value, especially to the Malays, whose interests in the past have been chiefly agricultural. Legislation requires miners to fill in and level excavated areas, to minimize discharges into the drainage system, and to stabilize mined-out areas. Conforming to such regulations has become more expensive with the mining of deeper tin deposits, which involve removing more tailings of waste material from the working area of dredges. A larger area, therefore, is necessary simply for the settlement and retention of the waste, and once the limit of gravity disposal is reached, pumping is required, which means more energy consumption.

Rates of Recovery and Tin Losses

Tin mining costs depend not only on the grade of ore, but also on the rate of recovery of tin from the ore, or, to put it another way, the loss of tin from the mining and treatment processes. An important way of maintaining profitability in the face of rising input costs and the burden of taxation, is through an increase in the rate of recovery or a reduction in the rate of loss of tin. The more tin that can be extracted from the material brought out of the mine, the better for the producer. From the point of view of the producing country and the world as a whole, the less tin left in a deposit after mining is finished the better for economic welfare, since future recovery may either be impossible or economically impracticable. It is, of course, impossible to achieve anything like 100 per cent recovery. No mining industry, whether fuel or non-fuel, can completely exhaust a mining area. Both technical and economic factors make it impossible to avoid leaving some ore or fuel in the ground. Moreover, as the Tin Council report points out, 'in the case of alluvial deposits certain areas will be lost to mining as they become overlaid by roads, new river beds, railways, buildings and crops'.[50]

The loss of tin from material dug out of lode or washed out of alluvial deposits varies greatly between producing areas. It is particularly high in the Bolivian underground mines, where the chemical and physical nature of the narrow veins makes separation of the cassiterite ore from the waste material extremely difficult. Some tin losses are unavoidable, others could be prevented by better techniques and methods. W. Fox comments on the position in the Bolivian mining region:

> Outside the mine, the techniques for treating the ore, particularly mixed ores, are sometimes lamentable. Much of the labour employed in handling broken ore is female and it is possible that in some cases a quarter of the tin won by the men with such labour is at once thrown away. Even where treatment, crushing, and flotation techniques are quite advanced, there may still be another loss of one third of the original ore mined'.[51]

With similar losses in the private Bolivian sector, the overall effect is that only about 50 per cent of the tin extracted from all COMIBOL and other mines is recovered.

The recovery rates for a number of treatment plants in several countries are listed in the Tin Council report. In Bolivia there is a range from 35 per cent to 80 per cent, with Catavi, by far the largest, having a 50 per cent recovery rate. The majority fall well below the maximum rate. The two large Cornish mines have recovery rates of around 70-75 per cent. The Renison mine in Tasmania, the world's largest tin mine, reported an overall recovery of 68.8 per cent in 1977-8, in spite of the complexity of the ore-bodies, and the 'difficulties experienced in achieving and maintaining an adequate rate'.[52]

Advances have been and are being made in recovery techniques which would considerably increase ore reserves by lowering the economic grade and make it worthwhile to treat the vast quantities of tailings which have accumulated over many years. There are expectations of a considerable improvement in Bolivia as the result of a new, expensive treatment plant, which should allow a higher recovery rate than is customary in Bolivia. As the Tin Council report points out: 'a 20 per cent improvement in recovery rates from the levels now considered acceptable by efficient hard-rock miners using current technology would add approximately 10,000 tonnes of tin metal to world supply annually without the addition of a single new mine'.[53]

Notes

1. P.W. Cullen and G.C.R. McDonald, 'Problems resulting from the Development of a Large-Capacity Mineral Recovery Dredge', a paper presented to the Fourth World Conference on Tin, Kuala Lumpur, 1974. Vol. 4 of the *Conference Proceedings*, London, 1974, p. 243.

2. ITC, *Tin Production and Investment*, p. 57.

3. ITC, op. cit., p. 71. 'An interesting new departure in Malaysia has been the development of very large and complex gravel pump mines, working to depths of 200 feet or more. Each of these mines is capable of shifting over 300,000 cubic yards of ore and overburden a month and of producing around 1,000 tonnes of tin-in-concentrates a year. They enable gravel pumping to be carried out in places and depth hitherto inaccessible to gravel pump mines and where dredging is not the ideal method.'

4. John S. Carman, *Obstacles to Mineral Development: A Pragmatic View*, Bension Varon (ed.), Pergamon Policy Studies, New York, 1979.

5. John Thoburn, *Primary Commodity Exports and Economic Development: Theory, Evidence and a Study of Malaysia*, John Wiley and Sons, London, 1977, p. 107.

6. *Tin International*, April 1977. Quoted in the third of a series of articles on the Indonesian tin mining industry, 'Indonesia looks to the Sea'.

7. Thoburn, *Multinationals, Mining and Development: a Study of the Tin Industry*, Gower, Farnborough, 1981, p. 79.

8. For information on the Renison mine the writer is indebted to correspondence with Bernard C. Engel, ITC Deputy Buffer Stock Manager.

9. W. Fox, *Tin: the Working of a Commodity Agreement*, Mining Journal Books, London, 1974, p. 41.

10. Fox, op. cit., p. 34.

11. Thoburn, op. cit., p. 68.

12. Thoburn, op. cit., esp. Chap. 6. See also his article, 'Commodity Prices and Appropriate Technology: Some Lessons from Tin Mining', *Journal of Development Studies*, Jan. 1978, pp. 36–49.

13. Thoburn, 'Commodity Prices and Appropriate Technology: Some Lessons from Tin Mining', p. 47.

14. Thoburn, op. cit., p. 48.

15. ITC, *Tin Production and Investment*, 1979. This section draws heavily on the material assembled in this report by H.W. Allen, Special Adviser to the Tin Council, and Bernard C. Engel, Deputy Buffer Stock Manager.

16. The different types of economic rent are discussed in the World Bank Staff Working Paper No. 354, 1979, *Development Problems of Mineral Exporting Countries*, pp. 86–93.

17. See Helen Hughes and Shamsher Singh, 'Economic rent: incidence in selected metals and minerals', *Resources Policy*, Vol. 4, No. 2, 1978. The distribution of rent for the commodities listed is estimated by their study as follows:

	Bauxite	Iron ore	Phosphate rock	Tin	Petroleum
Exporting country					
1972	36	49	14	90	15
1973	–	53	13	78	17
1974	93	61	78	61	47
1975	68	21	49	92	44
1976	50	40	27	85	44
Importing country					
1972	64	51	86	10	85
1973	–	47	87	22	83
1974	7	39	22	39	53
1975	32	79	51	8	56
1976	50	60	73	15	56

Helen Hughes also discusses the question of economic rent from minerals in 'Economic rents, the distribution of gains from mineral exploitation and mineral development policy', *World Development*, Vol. 3, 1975, pp. 811-25.

18. ITC, op. cit., p. 129.

19. ITC, op. cit., p. 129.

20. ITC, op. cit., p. 129.

21. It has been estimated that the amount of tin-in-concentrates smuggled out of Malaya to smelters in Singapore has been as high as 7 per cent of Malaysian production in the late seventies. See F.K.J. Jackson, 'Changing Patterns in Malaysian Mining', *Tin International*, June 1980.

22. ITC, op. cit., p. 150.

23. Thoburn discusses the taxation issue extensively in the works cited, also in 'Malaysia's Tin Supply Problems', *Resources Policy*, March 1978.

24. ITC, op. cit., p. 146.

25. Malcolm Gillis *et al., Taxation and Mining: Non-Fuel Minerals in Bolivia and other Countries*, Ballinger, Cambridge, Mass., 1978.
26. Gillis, op. cit., p. 241.
27. Gillis, op. cit., p. 240.
28. Gillis, op. cit., p. 291.
29. Knorr, *Tin under Control*, Food Research Institute, Stanford Univ., Calif., 1945, p. 86.
30. ITC, op. cit., p. 78.
31. Orris C. Herfindahl, *Copper Costs and Prices 1870-1957, Resources for the Future*, John Hopkins Press, Baltimore, 1958, p. 224.
32. Fox, op. cit., p. 6.
33. Much of the material available to the working party was included in the report written by the author for the Tin Council in 1964. See ITC, *Report on the World Tin Position, with Projections for 1965 and 1970*, London, 1965.
34. David J. Fox, 'The Bolivian Tin Mining Industry: Some Geographical and Economic Problems', a paper submitted to the Technical Conference on Tin, 1967, p. 11. See also David J. Fox, *Tin and the Bolivian Economy*, Latin American Publications Fund, London, 1970.
35. Herfindahl, op. cit., p. 230.
36. Thoburn, 'Commodity Prices and Appropriate Technology: Some Lessons from Tin Mining', *Journal of Development Studies*, Jan. 1978, p. 41.
37. Gillis, op. cit., p. 49.
38. ITC, *Tin Production and Investment*, p. 138.
39. Gillis, op. cit., p. 50.
40. Gillis, op. cit., p. 50.
41. Gillis, op. cit., p. 51.
42. Gillis, op. cit., p. 51.
43. Gillis, op. cit., p. 52.
44. ITC, *Notes on Tin*, No. 195, April 1977, quoting an article in the *Far Eastern Economic Review*, 1 April 1977.
45. ITC, *Notes on Tin*, op. cit.
46. ITC, *Notes on Tin*, op. cit.
47. ITC, *Tin Production and Investment*, p. 136.
48. ITC, op. cit., p. 136.
49. ITC, op. cit., p. 136.
50. ITC, op. cit., p. 88.
51. W. Fox, op. cit., p. 61. See also David J. Fox, *The Bolivian Tin Mining Industry: Some Geographical and Economic Problems*. There is a discussion of Bolivia's technical problems in Gillis, op. cit., pp. 54-6.
52. ITC, op. cit., p. 88.
53. ITC, op. cit., p. 88.

3 PRODUCTION TRENDS

It is a striking feature of the non-communist tin mining industry that, of the eight leading producing countries, four reached their highest output in peacetime before the Second World War, Malaysia and Indonesia in 1937, Bolivia and Nigeria in 1928. Zaire's peacetime peak came in 1953. Yet doubters about the future supply of tin, and of other non-renewable industrial materials, might take consolation from the fact that some present-day tin-producing areas have been in business for centuries. Cornwall was a tin producer before the Romans came to Britain. Malaya was certainly an exporter in the fifteenth century, and a producer in the ninth century. Indonesia was producing tin at the beginning of the eighteenth century. Thailand's history as a tin producer may go back 2,000 years.

Developed Countries

The United Kingdom

Growth of the Cornish Industry. Cornwall has a long history as a leading producer. Production was certainly small by modern world standards, but so was world consumption until well into the nineteenth century. By the middle of the century output had expanded substantially in Cornwall to meet a growing demand. Several hundred Cornish mines were then producing about 7,000 tonnes of tin metal, more than double the output of the other main producer, the Dutch East Indies. Cornwall did not reach its peak output of 11,000 tonnes until 1871. As late as 1893 output was still 9,000 tonnes, over 10 per cent of world production, but the UK was no longer self-sufficient in tin. Within three years it was down to 4,900 tonnes and exceeded 5,000 tonnes in only one subsequent year, namely 1913. The depression after the First World War reduced output to a few hundred tonnes, but during the boom conditions for tin in the twenties, output recovered to an average of 2,000 tonnes, with a post-war peak of 3,300 tonnes in 1929, the last good year for Cornwall before the war. From 1935 to 1938 the remaining small group of mines averaged 2,000 tonnes. Thereafter it was not until 1972 that the Cornish industry again reached this level.

The Effects of Foreign Competition. The decline of the Cornish industry, which was economically and socially painful for the Duchy, reflected its inability to compete with the rapidly expanding lower-cost producers in Malaya and Bolivia. In the last quarter of the nineteenth century Malaya had the advantage not only of inherently easier alluvial deposits, but also, from 1879, of a progressive depreciation of the Malayan silver dollar, the European value of which fell from 3s 8d in 1879 to 2s 1d in 1894. The Malayan workers were paid in silver, but the tin was sold for gold on the world market. D.B. Barton, a leading historian of the Cornish industry, comments on the drastic change in the ratio of value between the two precious metals: 'This was perhaps the greatest single factor in bringing Straits tin to its predominant position at this time, allied to the inherently easy mode of working alluvial ground'.[1] The International Tin Council *Statistical Yearbook* includes among the causes of Cornwall's decline in the last quarter of the century faults within the Cornish industry itself: 'the reckless and improvident method of financing and division of profits pursued by the nineteenth century companies, the reckless gutting of the richer ores in many mines, the unwillingness to provide capital for long-term development work, the dead weight of pumping, and the exhaustion of the better lodes'.[2]

The decline in the value of silver was also partly responsible for the growth of competition from Bolivia by pushing silver miners into the mining of tin. After several decades of little change in tin output, Bolivia expanded rapidly from the mid-nineties, assisted by a large fall in transport costs following the completion of a rail connection to the Pacific. Although working conditions in the mines were extremely difficult, their tin deposits were much richer than the Cornish ones. At that time, the Bolivian producers could discard as worthless anything except the richest ores. In fact, the waste dumps and tailings of the old silver mines were as rich in tin on average as some of the best Cornish mines.

The Cornish revival. The first signs of a Cornish revival after the Second World War came in the sixties, after the sharp increase in the price of tin. Prospecting began at the Wheal Jane mine in 1966. Wheal Jane had had a long chequered history. It was reopened in 1850 after a period of inactivity, then closed down in 1884 as competition increased from southeast Asia. For a few years before the First World War it was again in operation, then closed down in 1914. After remaining abandoned for over half a century, improvements in flotation techniques, which enabled the recovery of tin previously regarded as uneconomic, led to its reopening in 1971, the first new Cornish mine in 60 years. Wheal Jane was followed in 1976 by the opening of the Mount Wellington mine after eight years of prospecting and development at a cost of £6.5 million. The expected output of the Mount Wellington mine was 1,600 tonnes a year, similar to that of Wheal Jane. If everything had gone according to plan, the two new mines, plus the existing mines, would have met over 40 per cent of a diminished UK consumption.

The economic hazards of restarting deep underground mines in an old mining area were soon revealed. Mount Wellington closed down after producing less than 700 tonnes of tin in two years' working. Explaining the failure, the company stated: 'The tonnage of reserves and the average grade available were substantially below the projection on the basis of which the mine had been opened.'[3] Underground water problems and too low a tin price were also blamed for the closure. The failure of this project soon led to problems with the adjoining Wheal Jane mine, which also shut down, except for pumping operations financed by government money.

That all hope of a Cornish revival was not abandoned as a result of the débâcle, was shown by a new project under a different company, involving the same two mines. Wheal Jane resumed its chequered career in 1980, to be followed by Mount Wellington in 1981, their expected joint output being 1,500 tonnes. Assuming that the previous experience is not repeated, as much as one-third of the UK tin consumption could be met by the Cornish industry in the eighties. If the project were as successful as the existing Cornish mines have been for many years, the ghost of failure which has often bedevilled the Cornish might be laid. Writing in the mid-sixties, however, D.B. Barton took a sober view of the prospects for a successful expansion of tin mining in Cornwall:

> The resuscitation of an old, deep mining field is a difficult task – one, moreover, that has never yet been accomplished in the history of metal mining – and it is as fraught with the possibilities of expensive failure as is the resuscitation of a single old deep mine. The more ancient, the deeper, and the more intensive has been working in the past, the more remote are the chances of a successful re-working. It took neither money nor skill many years ago to open a virgin mine that outcropped at surface; at a re-working it required a little more outlay and mining ability to go down below water level . . . at a second re-working, at greater depth, with

> steam power necessary, de-watering and re-equipping involved the provision of capital and capable management; third, fourth, fifth, and successive re-workings merely increased the amount to be staked before there was even a chance of success, whilst at the same time the prize to be won had not increased and may well have diminished. Such was, and is, the gamble of re-opening metalliferous mines. Herein lies the central dilemma of Cornish tin mining revival – that there is still a great deal of tin lying scattered in the country, but that even using diamond drills as a means of insurance, a great deal of money can be lost trying to locate and extract it. The history of mining in Cornwall in the twentieth century shows this only too clearly.[4]

Whether this judgement is still valid in the eighties remains to be seen.[5]

Australia

The Nineteenth-century Mining Boom. The revival of tin mining in Australia, although very modest by the standard of other metals, is one of the success stories of the last 20 years in the mining industry. Once tin had been found in commercial quantities in the middle of the nineteenth century, mining developed with great speed. Average annual output in the period 1861-70 was only 130 tonnes. In the next two decades it averaged 8,600 tonnes and 10,000 tonnes. After this short period of growth the Australian industry felt increasing pressure from low-cost producers in southeast Asia, and output fell to an average of less than 6,000 tonnes in the decade 1891-1900. For a few years higher prices led to a small increase in output up to the First World War, but the exhaustion of known high-grade reserves and the fact that the miners could not yet deal with large volumes of lower-grade tin-bearing ground cut output after the war to around 1,500-3,000 tonnes.

The Modern Australian Industry. The expansion after the Second World War did not begin until the sixties. By 1967, Australia, the only industrialized country with a large tin mining industry, had become a net exporter of tin. Production in 1978-9 was over three times greater than consumption, and about 6 per cent of non-communist world production. It was now equal to the peak output during the first Australian tin mining boom of the seventies and eighties. Records of output before 1900 are probably incomplete for several states, but the peak may have been around 13,000 tonnes in 1881. For eight years up to 1881 Australia was the world's largest producer, accounting for up to 25 per cent of world production.

Most Australian production comes from underground and opencast mines, but there is also a small amount of dredging. As in the nineteenth century, Tasmania is the chief producing state. Tasmania has the world's largest tin mine with an output of over 5,000 tonnes annually in the late seventies. The rest of Australian output comes from less than a dozen large mines and another 250 or so small mines which were originally the backbone of the industry. These small units still manage to survive in a country characterized by large-scale, capital-intensive mining.

The recovery of the industry in the sixties depended only partly on prices. Technological improvements also played an important role by raising the previously uneconomic deposits above the margin of profitability. According to one authority, the development of new earth-moving equipment and new water supply schemes has helped to contain production costs within reasonable limits, and improvements in metallurgy and ore-dressing techniques have improved the recovery of tin concentrates of a saleable grade. Further improvements in technology would reinforce the strength of the Australian industry as one of the more reliable sources of tin. Prospects for further expansion, according to the Tin Council report, depend on

> success in prospecting for large-scale lower-grade deposits and on the progress achieved at a number of properties which have been known about for some time, but which have not proceeded to the development stage, due either to the sub-marginal nature of the deposits at historic tin prices or to the difficulties inherent in mining in remote areas of the country.[6]

Developing Countries

Malaysia

Development and Structural Change. Malaysia (formerly the Federated Malay States as part of the British Empire) has been the largest tin producer since 1883. Most Malaysian tin comes from two states, Perak and Selangor, and is found in extensive alluvial deposits, occurring at depths varying from a few feet to over 50 feet. A large new deposit to be mined in the eighties lies below 150 feet of overburden, involving dredging to a new maximum depth. Lode deposits are believed to exist in mountainous areas which may be mined in the more distant future, and would require different mining techniques from those appropriate to alluvial deposits.[7]

Before the expansion of production in Bolivia and Indonesia at the end of the last century, Malaysia had a 50 per cent share of world production. Under normal peacetime conditions the Malaysian share has never been less than 30 per cent since the 1880s. After the last war the industry recovered very quickly in spite of the security problems caused by the 'Emergency'. The 1938–9 level of output was regained by 1948, facilitated by the excess capacity which had existed in the dredging sector since the thirties and a speedy programme of reconstruction.

The share of the dredges remained higher for some years than in the thirties, reaching as much as 53 per cent in 1961 during the period of export control. Export control, in fact, hit the dredging sector less hard than the gravel pump mines, whose small output made severe cuts difficult to reconcile with profitable working. From 1961, however, the share of the dredges fell steadily to 31 per cent in 1968, around which level it has fluctuated moderately. The number of dredges did not change much in the sixties, but fell in the seventies to 53 in 1978, less than half the number before the war. The number of gravel pumps rose to over 1,000 at their peak in 1966, well above their prewar numbers, and in spite of a marked decline since the early seventies, it is still above the prewar peak. Between 1962 and 1968 the output of the gravel pump mines doubled and their share has remained consistently above 50 per cent since 1966.

A striking feature of the Malaysian industry in the seventies has been the relatively high level of output from dulang washing, the simplest possible method of winning tin. Since 1972 total output of the dulang washers has been consistently above that of all previous years, but total Malaysian output, delivered to the smelters at Penang and Butterworth (i.e. excluding smuggled concentrates) has not yet exceeded the 1937 peak of 77,266 long tons. It was, however, over 70,000 tonnes in every year from 1968 to 1973. Since 1974 the amount delivered to the smelters has not exceeded 65,000 tonnes. Some smuggling is believed to have been going on for a number of years.[8] Tin Council figures for southeast Asian production from 1975 include a sharply increased estimate for production of unspecified origin, which probably indicates smuggling. It seems unlikely that much of the unspecified tin could have come from the smaller producers like Burma, since there is no evidence that their output suddenly spurted in the second half of the seventies. It would not be surprising if some Malaysian output was smuggled out of the country to evade high export duties.

Problems in the Seventies. The failure of Malaysian output to expand in the seventies has been disturbing to consumers and can only confirm the belief of large American users that continued research into substitutes is essential. Fox has suggested that Malaysia has never shown that it was fully aware of the need to convince users that they could always rely on a regular supply of tin at what they regarded as reasonable prices.[9] It is clearly a controversial issue. Various reasons have been given for the decline in output in the seventies, including land problems and high taxation. Thoburn states that 'private prospecting effectively came to a standstill'.[10] Some prospecting seems to have gone on, but it is difficult to quantify. That prospecting could be highly successful in finding new large deposits was shown by the discovery of the Kuala Langat deposit in 1972, but reaching the actual production stage has proved to be a daunting obstacle race. After the initial agreement between the foreign mining company which found the deposit and the Selangor state government in 1975, it took five years to produce a formula for working the deposit. It was reported in 1980 that it would be another five years before the deposit produced any tin and it was still necessary to make a one-year feasibility study because of the problems in mining at a much greater depth than normal for an alluvial deposit.

Indonesia

Growth and Decline. Indonesia (formerly the Dutch East Indies) is estimated to have been producing about 1,000 tonnes a year in the middle of the eighteenth century. In 1850 production was about 3,500 tonnes. The subsequent expansion of the industry was more or less continuous up to 1929, but not nearly as spectacular as that of Malaysia. In the 1850s Indonesian output was only slightly below that of Malaysia, but by the end of the century it had dropped back to one-third.

Unlike Malaysia, there was no flood of British capital into the industry, nor was there a large number of independent Chinese miners such as made a major contribution to the expansion of Malaysian output. Indonesia was unique in the early participation of government in tin mining. In 1816 the Netherlands Indies government took over the mines on the island of Bangka, which had always been the main source of Indonesian tin. The government also had a majority holding in the company which operated the mines on the island of Belitung, where production started in 1852. Most Indonesian tin comes from the alluvial and offshore deposits of Bangka and Belitung. There is only one deep primary deposit which produced up to 2,000 tons a year until the underground mine was flooded during the war, and has so far defied efforts to make it an economic proposition again by dewatering.

Dredging started later than in Malaysia. The first dredges were not operating until after the First World War, but Indonesia followed the lead of Thailand in offshore mining and quickly became the leading offshore producer. By 1939 there were 22 dredges onshore and offshore, producing about two-fifths of total Indonesia output. In 1937 the industry reached its prewar peak, but it had substantial spare capacity in the dredging sector which was not fully used until the brakes were taken off output in 1940–1. Production in 1941 shot up to 53,000 tonnes, about 45 per cent above the prewar peak. A similar sharp increase occurred in 1948 when the industry was recovering from the war and the US was buying heavily for the strategic stockpile. By 1954 output had risen to 36,000 tonnes, only slightly below the 1937 peacetime peak, but this proved to be the maximum up to the time of writing. From 1954 production fell in every year until 1963. There was a short-lived recovery in 1964–5, then a further fall to the nadir of the nationalised industry in 1966, where output was the lowest this century. Since 1966 there has been a gradual, more or less continuous recovery, but output in 1979 was still more than 15 per cent below the 1954 level.

The precipitous decline was the result of a number of factors, both economic and political. W. Fox has summed up the situation of the nationalized industry in the early sixties:

> It suffered from almost every disease that could affect an industry – from the continued depreciation of ageing dredges, from the misbuying of equipment, from gross inefficiency in management and from the political interference to be expected from an industry which was still earning for Indonesia the equivalent of around £16 million in hard currency in its worst years.[11]

Fox goes on to comment that handing over the running of the mines to the army in 1963 did not raise efficiency, 'except in the art of smuggling'. A major factor, possibly, in the last stages of the decline was the loss of Chinese workers after their expulsion from the tin producing islands in the great pogrom following the abortive communist *coup* in 1962. With such a catalogue of misfortunes it is hardly surprising that output fell drastically. An even greater contraction might have been expected.

Problems of Recovery. The very gradual recovery since 1966 is in marked contrast to the speed of recovery after the dislocation caused by the Japanese occupation. It is difficult to escape the conclusion that the expulsion of the Dutch company in 1953 and 1958 left problems which have had long-lasting effects on the industry. Foreign enterprise would probably have led to a much more rapid expansion if it had been acceptable to the government. This is suggested by the experience of an Australian company which entered into a joint venture with the state corporation in 1971. Exploration began in 1971 and pilot plant production in 1973 was 152 tonnes, rising yearly to 2,900 tonnes in 1978 and 3,800 tonnes in 1979. It might also be argued that a greater employment of expatriates would have led to a faster rate of recovery, particularly in the early stage. Nevertheless, it seems that a big expansion of output could have been achieved only by the modernization and expansion of the industry's capital stock. Even now, in 1982, two-thirds of the dredge fleet is 40 or more years old. Of 30 dredges built between 1926 and 1978, 9 were built before 1930, 12 between 1930 and 1940, 7 in 1947, 1 in 1966 and 1 in 1978. The main increase in output is due to investment in offshore dredging. Between 1971 and 1979 offshore dredges raised their output from 3,624 tonnes to 11,845 tonnes, whereas production by the onshore dredges fell from 4,024 tonnes to 3,517 tonnes. Gravel pump output rose from 5,963 tonnes to 12,932 tonnes.

Some of the problems of the Indonesian industry arise from difficulties over 'back-up services'. The more isolated mines suffer from their remoteness from supplies of spare parts and from high transport costs. The nationalized corporation tries to overcome the problem of spare parts and maintenance by carrying a large stock, but, according to a spokesman, this involves locking up a large amount of capital which might be used for investment. There is a marked contrast with Malaysia where many small outside workships meet the industry's needs locally. It is suggested, however, that government regulations in Indonesia are 'not conducive to attracting private back-up services'.[12]

Thailand

Growth of Tin Mining. Of all the larger producing countries, Thailand seems to be the most promising, given reasonably stable internal conditions. Thailand became the second largest producer in 1979, when its output of 33,962 tonnes was slightly higher than that achieved by any other country, except Malaysia, since 1954.

Thailand's tin deposits are part of the extensive tin belt which stretches into China and south to the tin islands of Indonesia. As in Malaysia, tin is mined chiefly by dredges and gravel pumps, but in contrast to Malaysia, Thailand has long had a substantial offshore dredging sector, to which was added in the seventies suction boat mining. There is only a small amount of underground mining, not reported separately by the Tin Council.[13]

Like other parts of southeast Asia, Thailand has a very long history of tin mining, possibly going back over 2,000 years. Tin was an important article of Thailand's trade with China, later with Arabia, and, through the Portuguese, with Europe in the sixteenth century. The birth of the modern industry is attributed to Chinese mining of placer deposits in southern Thailand in the first half of the nineteenth century. By the 1870s Thailand's output was averaging about 2,800 tonnes, but there was no rapid increase during the last quarter of the century when other countries were experiencing high rates of growth. In the years immediately preceding the First World War, output averaged 4,500 tonnes. The war led to a sharp increase to over 8,000 tonnes, but production slipped back in the twenties and did not exceed the wartime peak until 1929.

Thailand was much less affected than other important producers by both the depression and the restrictions on output in the thirties. By 1937 production was more than double the average during the twenties. Thailand took longer to recover after the war than other southeast Asian producers. The lowest level of the century was reached in the early postwar years when other countries were well on the way to recovery. In the first phase after the war, Thailand reached 10,500 tonnes in 1950, but it was not until the sixties that a major expansion began, coinciding with the decline in Indonesia, Bolivia and the newly independent Congo. By 1965 the previous peak had been exceeded and in 1968 output was a record 23,980 tonnes. After 1968 there was no further expansion until the late seventies, fortunately at a time when Malaysian production was stagnating.

Structural Change. The expansion in the seventies has been associated with the appearance of the suction boat operators in offshore waters hitherto worked by dredges. It has been estimated that 3,000 small converted fishing boats have been using the suction boat method of lifting tin-bearing mud from the seabed. Both the Tin Council and the Thailand Offshore Mining Organization (OMO) have been strongly critical of the activities of the suction boats.[14] The Thailand OMO believed that because of their haphazard methods and the poor rates of recovery the actual loss of tin was greater than the 3,500 tonnes of tin which it estimated were produced by this method from OMO leases in 1975. All of this output was smuggled abroad for smelting. Moreover, at one stage foreign-owned dredging companies were literally pushed out by the illegal miners. After several years of illegal operation, the suction boat operators were eventually legalized by the government, acting apparently under the old principle, if you can't beat them, join 'em. Since their legalization in 1976, the share of the suction boats in Thailand's total output has risen to one-third in 1978 and nearly 40 per cent in 1979.[15]

Before the war, dredges accounted for about two-thirds of Thailand's output. In the fifties and sixties their share fell steadily until by the early seventies it was down to one-quarter compared with 60 per cent in the early fifties. Gravel pump mining became much more important as new power-driven machinery was introduced and old-fashioned recovery methods replaced. Between 1950 and 1972 the number of non-dredging mines doubled. Another large increase in 1977-9 increased the number of gravel and hydraulicing mines to 386 compared with 281 in 1972 and less than 70 in the early fifties. There is no doubt, therefore, that the entrepreneurial drive in Thailand to produce more tin, given the right incentives, was highly successful. Local enterprise has shown a vitality and flexibility often less manifest in developing countries. The elasticity of supply has sometimes been high, and most of the increased output has come from indigenous miners, notably the enterprising, if somewhat freebooting, suction boat operators.

Bolivia

Growth and Structure. As a tin producer, Bolivia has a much shorter history than the southeast Asian countries. Tin became important only as silver mining lost its appeal to Bolivian miners in the second half of last century. The Tin Council reports an average tin production of only

100 tonnes in the 1860s. Production still averaged less than 1,000 tonnes in the 1880s, less than 2 per cent of world production, and well below production in the now shrinking Cornish industry. By the end of the century the Bolivian share had risen to 5 per cent, and was approaching 20 per cent just before the First World War. The most buoyant phase of the industry came in the second half of the twenties when production rose by about 50 per cent, from an average of 29,500 tonnes to 46,500 tonnes in 1929. Production was severely cut in the worst years of the depression, then it remained well below the levels of the late twenties as a result of serious production problems.[16] The removal of checks to production during the war, with the US buying heavily, led to a sharp increase in output, which exceeded 40,000 tonnes in 1941, 1943 and 1945, but never reached the 1929 peak, which has remained above all subsequent outputs. The best year in the seventies did not greatly exceed two-thirds of the peak.

The Bolivian mining region is unique, not only in the world tin mining industry, but in world mining generally, by reason of its location and the difficult conditions under which the miners must work. The mines are in the Andes at heights above those of the highest mountains in Europe. Twenty-five of the largest mines are more than 11,500 feet above sea level. Several are more than 15,000 feet above sea level. The administrative headquarters of the biggest mining complex are at 12,700 feet. The entire mining community lives under conditions which are difficult to imagine in Western Europe. The mining scene has been described by W. Fox, a sympathetic observer of the Bolivian miner's way of life:

> Mountain heights, and all that comes from them, are the norm. The landscape is arid, bleak, cold, sometimes very beautiful and always without humanity. The air is thin; heavy physical labour is difficult. The hundreds of mines are disconnected; roads are poor, sometimes little more than tracks . . . The possibilities for decent agriculture are limited; there is almost no industrial manufacturing development within reach of the mining townships.[17]

Most Bolivian tin comes from underground mines, but a small dredging sector has been developed in recent years, producing as much as 2,290 tonnes in 1977, falling to 1,415 tonnes in 1979. Since 1976 the share of the dredges has been around 5-7 per cent of Bolivian output. The nationalized sector includes some of the world's largest tin mines. The medium-sized mines sometimes produce as much as the average southeast Asian dredge, but many small mines may have an output of 2 tonnes or less. The working of many of the small mines has been picturesquely described by Fox, showing the striking contrast between the different types of underground mines:

> Close to the largest and one of the most highly mechanised tin mines in the world, a small unit of four men drives its own shaft down to about 100 feet, instals a windlass, sends down one man to hew out the ore in the immediate underground standing space, winds up the broken ore in a bucket, prays to God and lights candles to the Devil for no cave-in.[18]

Production Problems and Mining Costs. Ore grades are much lower in the Bolivian mines than they were in the 1890s when production was switching from silver to tin. Fox states that the fortunes of the Patino family originated partly from ores of no less than 47 per cent tin content, admittedly exceptional, but 8–12 per cent tin was common at the time.While grades fell very sharply in the thirties, this did not prevent the big increase in output during the forties when there was a market tendency to boost output by the wasteful practice of 'picking out the eyes'. As the previous chapter has shown, there was a big fall in grades after the Second World War, accentuating the many difficulties facing especially the state mining corporation.

Generally speaking, the Bolivian industry is the highest cost part of the world industry, although there are considerable variations, especially between the state corporation and some privately owned mines, which are more economical in the use of labour. The relatively high costs of the Bolivian industry are the result of a number of factors, some inherent in the nature of the industry, determined by the facts of geology and geography, others man-made and presumably capable of change.[19] They can be listed as follows: first, narrow vein deposits; second, a fall in grades, which were at one time high enough to compensate for difficult mining conditions; third, the complexity of the ores, with consequent difficulties for the processes of concentration and smelting; fourth, losses of by-products in the concentration process; fifth, high transport costs in both bringing in supplies and shipping out concentrates; sixth, problems of organization in the nationalized corporation; seventh, persistent, occasionally violent labour problems; and finally, lack of investment money. It is a formidable list of obstacles to the running of an efficient, competitive industry whose contribution to the national economy is essential.

Problems of the State Mining Corporation. The state corporation ran into severe difficulties in the early sixties, difficulties which had been masked for some time by the Tin Council's export controls. How badly the industry had been affected by accumulating problems in the fifties, was shown by the behaviour of output when export control came to an end. The output of COMIBOL, the state corporation, in 1962–5 averaged less than half the 1953 level. Other mines showed clear signs of recovery. Since then, the medium mines have maintained a more or less consistent level of output, well above that of the fifties. COMIBOL's output rose gradually to a post-nationalization peak of 23,306 tonnes in 1977, about 95 per cent of the 1952 level, then fell to 19,000 tonnes in 1979, with a further fall in 1980.

COMIBOL's crisis in the early sixties was tackled by the so-called 'Operation Triangle', under which substantial loans amounting to US$38 million were granted by the US, Federal Germany and the Inter-American Development Bank for the rehabilitation of the industry. The subsequent history of production showed that only a partial recovery was achieved. More investment was required, particularly for exploration and development, since the industry's problems could be traced back well before nationalization in 1952, and derived from insufficient investment in the thirties and forties. According to the Gillis study, 'reinvestment was minimal in the industry after the mid-thirties',[20] with no significant improvement after nationalization of the larger mines. 'Exploitation and development in and around COMIBOL's Catavi mines, potentially a rich source of new deposits, have been stagnant for years owing mainly, it is said, to the additional risks of investing in a mining zone beset with continuing labour difficulties'.[21]

Labour difficulties have often been referred to as a source of weakness in the running of COMIBOL, the product of revolution after many years of bad labour relations under the 'tin barons'. COMIBOL is generally believed to be over-manned, although this may not be true of underground labour, which has a high rate of wastage because of the arduous working conditions. Gillis says of the large Catavi mining complex:

> Partly due to inadequate maintenance of the mostly half-century infrastructure of the mine, and partly due to poor supervision and hesitancy to invest in a mining area that has long been the centre of labour union militancy, the rate of fatality and dismemberment for COMIBOL labourers who work underground is far higher than for

most of the larger medium mining firms. Health and work conditions in this group of mines are in general poor. Lack of both ventilation and water for drilling, and the resultant high dust levels, contribute to high rates of silicosis. As a result, workers in Catavi may expect an average of from 5 to 7 years of underground work. This compares with an average underground work expectancy of 15–20 years in some of the better-ventilated private medium mines working vein deposits.[22]

Brazil

Brazil is reported to have been mining some tin early in the century, but production on a significant commercial scale did not begin until the late fifties. The Tin Council's statistics give only about 200 tonnes annually until 1958 when the first signs of expansion appeared. Production grew very slowly during the sixties and early seventies. In 1974–5 it averaged only 3,800 tonnes compared with 1,700 tonnes in 1964–5, but by the late seventies it was approaching 7,000 tonnes, making Brazil the sixth largest producer, following the decline in Nigeria and Zaire. The growth of production did not leave much scope for exports, since Brazilian consumption had begun to increase, particularly with the expansion of the tinplate industry.

Brazilian tin deposits are alluvial, in contrast to those of the Bolivian tin belt in the High Andes, and bear no relation to the Bolivian mountain lodes which are some 780 miles away. Tin has been found in a number of areas in the Upper Amazonian Basin, as remote from the coast as the deposits in Zaire. Apart from a small number of dredges, gravel pumping is the main method of mining, as it is in southeast Asia. Grades of ore are believed to be relatively good. According to Fox, some of the early discoveries had phenomenally rich grades of up to 200 kg per cubic metre, compared with an economic cutoff grade of 0.6 kg.[23] Most significant deposits probably lie within the more normal range of grades, but there is still a lack of information about the Brazilian mining areas.

During the sixties much of Brazilian output came from the very small operations of the 'garimpeiros', a Brazilian term describing independent prospectors and pick-and-shovel miners, who worked with the minimum of equipment and used extremely wasteful techniques, mining only the richest grades and possibly damaging the economic prospects of some deposits in the long run. After the banning of such mining activity by the government in 1971, the way was clear for more efficient operations by organized companies using modern mining techniques.

Nigeria

Early Production and Techniques. The British found indigenous miners working small alluvial deposits in 1884, but did not begin production themselves until 1909, after which the industry expanded gradually to its prewar peak of 11,118 tonnes in 1929. Nigeria, like other safe producing countries during the last war, was expected to increase output to the limit. It was found, however, that with limited spare capacity, the annual rate of output could be raised by only 1,000 to 1,500 tonnes from 1940 to 1944. Once the guaranteed market of the war period ended and costs began to press on profitability, output reverted to the level of the late thirties. Apart from the compulsory cuts during the period of export control in the late fifties and beginning of the sixties, the industry maintained a very stable level of production until 1968–9. Since then the history of the industry has been one of virtually unbroken decline. Production fell every year from 1968 to 1979. By 1979 it was at its lowest since before the First World War, a mere 2,750 tonnes, only slightly more than Cornish output in that year.

The Nigerian tin deposits are worked by surface mining, chiefly by gravel pumps and dragline methods. In the limited mining area the environmental effects of surface workings are particularly noticeable because of the need in the early days of expansion to cut down the forests for fuel. Environmental regulations have become more stringent since the sixties, a not unusual development in modern mining, and one which hits the highest-cost producers with most severity.

The Decline of Tin Mining. There are, however, more powerful reasons for the long decline of the Nigerian industry. The most easily accessible deposits have been exhausted, and it has not yet proved technically possible to work the expected substantial reserves of the sub-basaltic deposits in old riverbeds. Further, Nigerian mining methods were traditionally labour-intensive, which became a serious drawback when the general level of wages came under pressure from the growing effects of the booming oil industry on the national economy. Many mines had been over-lavish in their use of local labour. Rapid increases in wages in the late sixties and seventies put great strain on the cost structure of European-owned companies. Retrenchment following the wage increases and the government's policy of Nigerianization led to a drastic reduction in the number of expatriates employed, the cessation of mining in some areas, and an increase in contract labour. Contracting out mining had already been a feature of Nigerian mining, since it had been found that the local contractor was able to achieve higher productivity from his employees than was possible by his opposite number in the company. Retrenchment did not help the modernization of the industry, much of whose equipment was obsolete by the seventies. There also seem to have been problems in importing mining equipment and mining experts.

The lack of buoyancy in the Nigerian industry after the war had been foreseen by Knorr in 1944. His comments on the postwar prospects are illuminating: 'expanded production resulted from the accelerated exploitation of the richest cassiterite deposits, a practice which can scarcely be perpetuated ... the long-run capacity of the African tin mining industry may, therefore, prove to be below that attained in 1944'.[24] A more recent commentator, writing in 1971, foresaw a drastic decline in the industry as a result of cost inflation.[25] The record of the seventies certainly confirmed his gloomy forecast.

Zaire

The other leading African producer, Zaire, formerly the Belgian Congo, has shared the Nigerian decline. Excluding Brazil, Zaire is the most recent producer. The Congolese industry did not really expand until the mid-thirties when the established producers were restricting exports. The growth of production was rapid. From a mere 189 tonnes in 1931, production rose to nearly 10,000 tonnes in 1938. As a safe producing area, production in the Congo was increased to over 15,000 tonnes from 1943 to 1945. There was some decline in the early post-war years, but production was still over 13,000 tonnes as late as 1953-5. The troubles following independence in 1960 led to the first sharp decline, although after the removal of export control the industry succeeded in maintaining a stable rate of output of around 6,000-7,000 tonnes. After 1971 came the second phase of the decline, bringing production down steadily to 3,300 tonnes in 1979, the lowest since 1933.

The contraction in Zaire is not due to the exhaustion of reserves, which are believed to be good. The industry has suffered seriously from under-investment, especially since the troubles of the sixties. Difficulties in buying foreign equipment and materials because of foreign exchange problems, have handicapped producers.[26] Lack of skilled labour has played an important part in the decline. The general political and economic climate in the country has discouraged foreign companies. Costs have been inflated by acute transport problems, which have bedevilled an industry in the heart of the continent, in marked contrast to the better situated mining areas of southeast Asia.

The Centrally Planned Economies

The Soviet Union

Both the Soviet Union and China are large tin producers, but only estimates can be made in the West about their output. The US Department of the Interior estimated in 1980 that Soviet output had exceeded Bolivian output for the first time in 1978. Its estimate of 33,000 tonnes would make the Soviet Union the second largest producer in 1978. Previous estimates put Soviet output within the 25–30,000 tonne range. The Tin Council gives no figure for output, but reports from other sources that Soviet imports averaged 9,700 tonnes in 1975-6, rising to 13,900 tonnes in 1978–9.

Most Soviet tin is believed to come from underground mines, none of which are in European Russia. Little is publicly available in the West about the quality of the ores being mined. Some deposits were mined before and during the last war, but most discoveries seem to have been made since the war. The chief areas of tin mining are in the Transbaikal region, in the Vladivostock area north of the Amur river, and in the far northeast of Siberia, mainly north of the Arctic Circle. One American authority believes that the deposits are abundant and widespread in these areas.[27] An indication of the progress made in finding, and possibly opening up, new deposits has been the numerous technical papers published since 1945, which leads to the inference that there are many deposits and probably that many new mines have been started since the early fifties. The location of these deposits implies that mining and other costs could be very high. It has been suggested that the opening up of Soviet mines has been uneconomic, and that it would have been better to expand gold production in order to pay for increased tin imports.[28] This would, of course, conflict with the general Soviet policy of aiming at a high degree of self-sufficiency in industrial materials.

China

Estimates of Chinese production since the 40s are of doubtful accuracy, but a report suggests that more information is becoming available as Chinese relations improve with the West. According to this report, about 60 per cent of Chinese production comes from the Gejiu district of southern Yunnan, with a grade of 0.05 per cent tin, similar to that of Wheal Jane in Cornwall.[29] Like the Bolivian industry, the mines are in a mountainous area, but not up to the same high altitudes. Some are believed to occur at 8,000 feet.

Tin mining in China has a very long history, and as previous sections will have shown, Chinese miners have played a major role throughout southeast Asia. Output figures going back to the 1850s show a sharp increase at the same time as the expansion in the rest of the world industry. Production averaged 2,000 tonnes in the 1870s, and after some fluctuations grew to about 7,600 tonnes just before the First World War. Between 1916 and 1920 production averaged 9,600 tonnes. During the thirties, when the cartel was controlling exports, China, as an outsider, increased production to a peak of 13,400 tonnes. Since the communist regime was set up, the only statistics which give some solid indication of production are figures of imports into Western countries from China. From 1955 to 1960 the Soviet Union is reported to have imported Chinese tin at an average rate of nearly 15,000 tonnes. Known exports from China fluctuated between 2,000 tonnes and 13,000 tonnes in the seventies, the peak occurring in 1975 when a weak market forced the buffer stock to make its biggest purchase since the 1958 crisis. Estimates of Chinese production in the seventies have been around 22–23,000 tonnes, with not much chance of an increase in the near term until the industry is modernized.

Trends and Fluctuations in World Production

World tin production increased rapidly from the middle of the nineteenth century, averaging 18,000 tonnes in the years 1851 to 1860, and rising to an average of 98,500 tonnes in the years 1901 to 1910. There was a further increase up to the First World War, followed by a period of either decline or relatively slow growth. The 1913 output of 134,000 tonnes was exceeded only once up to 1924. Although production continued to grow in the early twenties after the immediate postwar slump, the rate of growth did not sufficiently match the growth of world demand to prevent a sharp increase in price. The sale of tin accumulated in the so-called Bandoeng Pool reinforced world supply, but was insufficient to prevent the price increase.

The first sign that a growing investment boom in the dredging sector was enabling output to catch up on demand came in 1927, when world production rose by 12 per cent in a single year. This rate of increase was repeated in 1928 and 1929, by which time production was 45 per cent higher than in 1913, a level of output not exceeded again until the best year of the thirties, 1937. From then, apart from the abnormal war years, 1940 and 1941, the prewar peak was not regained until

1972. The wartime peak has still not been exceeded, but it must be borne in mind that there was a desperate rush to produce as much tin as possible before the war spread to southeast Asia, and there was virtually unlimited demand from the US. Moreover, there was spare capacity in the industry after the investment boom of the twenties, and no government impediments to the response by private enterprise to an unheard-of buoyant market.

Production recovered quickly after the Second World War. Between 1946 and 1948 there was a remarkable increase of about 60 per cent, practically all of which was due to the two chief producing countries in southeast Asia, Malaysia and Indonesia. After a further increase of 10 per cent between 1948 and 1950, production settled down more or less on a plateau until 1958, when a recession and export controls combined to cut it by 30 per cent. The pre-recession level of output was not regained until 1966, even after restrictions were completely removed in 1961.

The slow recovery of output in the first half of the sixties, when consumption was about 20 per cent higher than in the fifties, was in marked contrast to the rapid recovery from wartime dislocation. The causes of the difficulties in the early sixties have been described in the previous account of individual producing countries, and can be summed up as follows. Political and economic problems in Indonesia, Zaire and Bolivia had a high cost in terms of lost output. If these countries had been as responsive as Malaysia after the removal of export control, they would have been able to produce substantially more, and less, if any, would have been required from the surplus US stockpile. Secondly, the severity of export control dislocated production and investment planning. Thirdly, American stockpile policy caused uncertainty among producers generally, especially in the more capital-intensive parts of the industry, although it does not seem to have inhibited new producers in Australia. Fourthly, there was no influx of foreign capital into the dredging sector in southeast Asia in response to higher demand and higher prices.

After reaching the pre-restriction level, output rose by about 10 per cent between 1966 and 1968, then fluctuated slightly around 185,000 tonnes until the late seventies, except for a 5 per cent increase in 1972. Before the next period of expansion starting in 1977, there was again a temporary shortage, compensated by heavy US stockpile sales, which did not prevent a sharp price increase. From 1977, output rose in two years to 200,900 tonnes, the highest peacetime level in the industry's history, in spite of a relatively low output in Malaysia and a more or less static output in Bolivia. This expansion was chiefly due to a sharp increase in Thailand.

Production and Prices

The price elasticity of supply is generally low in the short run, particularly in the absence of some spare capacity. Without spare capacity, the eventual response of output to a sustained rise in the market price, assuming market forces are allowed to operate, can be measured in years. Given freedom of investment, however, supply should adjust itself to a rise in demand, and the market price should normally settle down in accordance with the long-term trend, subject to the fluctuations which bedevil a commodity market.

Knorr's study of the prewar tin market emphasized the importance of the low supply elasticity for market fluctuations.[30] He argued that elasticity in the interwar years was lower than in the early years of the century because of the greater capital intensity of the industry. He found that once the great expansion of the dredging sector had occurred, and the gravel pumps had become more mechanized with an increased use of power, there was sufficient capacity by the start of the thirties to increase the supply elasticity even in the short run. With spare capacity and high overheads, there was a greater incentive in the thirties to increase output when the price fell on a weak market. It was to deal with such a situation that export controls and standard tonnages were used extensively in the thirties.

The effects of price changes on production in more recent years have been the subject of two inquiries, one by the World Bank, the other by the International Tin Council, the former using econometric techniques, the latter using the results of separate studies carried out by authorities in a number of producing countries.[31] The World Bank model covers the period from 1961 to 1975, which included a number of years in the sixties when normal economic responses were violently distorted by political events in three important producing countries. The World Bank team stresses that 'the diversity of production methods, industry structure, and the degree of government controls that exist in the various producing countries makes any general hypothesis about the producers highly tentative',[32] but it concludes that it was possible, even with such limitations, 'to capture most of the behaviour of the world tin economy with a small model consisting of 23 behavioural equations'.[33]

Table 3.1: Tin Supply Elasticities

Country or area	Price elasticity of supply	
	Short-term[a]	Long-term
Malaysia	0.31[b]	0.80[b]
Indonesia	0.21	0.91
Thailand	0.60[b]	1.25[b]
Bolivia	0.24[b]	1.34[b]
Developed countries	0.30[b]	0.70[b]
Rest of the world	1.11[b]	2.09[b]

Notes: a. Elasticities were estimated at the means of production and prices using the results of linear statistical models. b. Significant at the 95 per cent confidence level.

Source: World Bank Staff Commodity Paper No. 1, *The World Tin Economy: an Econometric Analysis*, June 1978.

The Bank's estimates of short-run and long-run supply elasticities for the four largest producers, derived from the model, are given in Table 3.1, as are supply elasticities for two groups of smaller producers, the developed countries and the rest of the world. It will be seen that there are considerable differences in elasticities between countries. Thailand emerges as the country with easily the highest short-run response, as well as a much higher response than either Indonesia or Malaysia in both the short-run and the long-run. This could be partly because Thailand, during the period covered by the study, was the only developing country to leave the industry relatively free from political interference, with the result that price and output tended to show the kind of relationship postulated by economic theory. There could be other explanations. The structure of the industry in each country could be relevant. Thailand produced a higher proportion of its tin from gravel pump mines than other countries during the period covered by the inquiry. Being much less capital-intensive than dredging or underground mining, gravel pump mines have a relatively quick response to a favourable price change provided the land is available.

The Bank's estimated long-run supply elasticities are higher than the short-run elasticities for all the countries, particularly for Bolivia, again with variations. 'The lag within which supply reacts to price changes varies among producers; Malaysia, Thailand, and the small producers (lumped together in the rest of the world) respond to price changes with a one-year lag; Bolivian, Indonesia, and the developed countries with a three-year lag'.[34] Average supply elasticity for the industry as a whole is put at 0.42 in the short run and 1.07 in the long run. Thus a 1 per cent rise in the real (i.e. deflated) price of tin would, according to this study, bring about a slightly more than 1 per cent increase in supply after a time lag of up to three years.

How much weight should be attached to these estimates of tin producers' response to price is debatable. The Tin Council report comments: 'Simplified coefficients of this kind are not very helpful in the analysis of factors underlying production response to price because elasticity would differ between methods of mining and because many influences other than price (e.g. non-commercial supplies) intervene, which have a bearing on medium- or long-term production movement'.[35] The report is obviously thinking of the complications arising from the uncertainties over the disposal of the surplus US stockpile, which must have had some effect on production and investment decisions.

The evidence on supply elasticities in the Tin Council's report was obtained in a different way from that used by the World Bank team. It was derived chiefly from model studies submitted by member countries on the theoretical cost of developing new mines of various types in these countries. The report gives information on the lead-times between the decision to invest in new capacity and the beginning of production. The estimates are made on certain assumptions which are not necessarily the same for each country. The report points out that 'an assumption basic to the Malaysian model studies is that no snag will be encountered in bringing the model mines into production', an assumption which might raise hollow laughter among the prospective developers of the Kuala project.[36] The report notes that a number of snags may hold up a project in spite of the most detailed planning; for example, difficulties over finance, shortages of equipment and protracted delays in obtaining prospecting and mining rights.

Subject to these provisos, the model studies indicated the following pre-production phases. From the start of exploration through mine development, the purchase of equipment, its installation on the site and the creation of the necessary infrastructure, it could be four to six years before an offshore dredging operation produced a single tonne of tin in Indonesia or Thailand. The Indonesian figure, with production starting up in the fifth year, was based on actual commercial ventures. A Malaysian onshore dredge would be expected, according to a model study, to start up in the sixth year, like a Thai offshore dredge. As expected, gravel pump mines have a much shorter gestation period. The Thai model showed a quicker response than the Malaysian, with production in the second year, as against the third year. Underground mines naturally had a long gestation period. The Australian example, derived from a specific commercial venture, showed production starting in the seventh year; the Bolivian model gave the fifth year.

Once again it should be stressed that the comparability of the results is affected by differences in the assumptions made by the reporting agencies. According to the report, differences in lead-times 'by no means indicate that countries in which a longer time is taken for a mine using the same method to come into production are less efficient'.[37] What emerges from this picture of production planning is that, unless there is some spare capacity, say, idle dredges or gravel pumps, or the possibility of switching to better grades, or sizeable stocks, sudden increases in demand can mean a lengthy period of relatively high prices.

Notes

1. D.B. Barton, A History of Tin Mining and Smelting in Cornwall, Barton, Truro, England, 1967, p. 214. For the early history of the Cornish industry see also the series of articles by Anthony Smith in Tin International, various issues in 1980 and 1981.
2. ITC *Statistical Yearbook*, 1962, p. 174.
3. *Tin International*, May, 1978.
4. Barton, op. cit., p. 287.
5. It was reported that a small pilot project involving suction dredging about one-quarter of a mile offshore was under way in Cornwall in 1980. If successful, the project could reach about 1,000 tonnes per annum eventually, according to the report in *Tin International*, Nov. 1980.
6. ITC, *Tin Production and Investment*, 1981, p. 38.
7. A detailed history of tin mining in Malaya has been written by Yip Yat Hoong, *The Development of the Tin Mining Industry in Malaya*, University of Malaya Press, Kuala Lumpur, 1969.
8. *The Far Eastern Economic Review*, 1 April 1977, quoted in ITC, *Notes on Tin*, April 1977. See also F.K.J. Jackson, 'Changing Patterns in Malaysian Mining', *Tin International*, June 1980.
9. W. Fox, op. cit., p. 24.
10. John Thoburn, 'Malaysia's Tin Supply Problems', *Resources Policy*, March 1978, p. 34.
11. W. Fox, op. cit., p. 35.
12. 'Indonesia Looks to the Sea', *Tin International*, March 1979. Articles on the Indonesian industry also appear in the February and April issues.
13. ITC, *Tin Production and Investment*, pp. 25–6. See also the report on Thailand by Paul F. Scholla and Ass., *Tin Deposits of Thailand*. According to the Scholla report, 'offshore deposits offer good possibilities, especially those employing modern dredges and new methods of recovery in depths beyond the capacity of present bucket or suction dredges'. The report also states that 16 tin-bearing granite ranges offer a large area for the exploration of lode deposits.
14. ITC, op. cit., p. 57.
15. Thoburn suggests that the tin lost by suction boat mining 'could easily be won if the area were re-worked by other means at some later date', since the offshore situation is unlike that of underground mining, where low-grade, once by-passed, may be lost for ever. See Thoburn, *Multinationals, Mining and*

Development, Gower, Farnborough, 1981, p. 137.

16. Bolivian output was badly affected by the loss of skilled men in the Chaco war. For several years Bolivia could not fulfil its export quota. See Fox, op. cit., p. 162, for a discussion of Bolivia's difficulties at that time.

17. Fox, op. cit., p. 60.

18. Fox, op. cit., p. 67.

19. Malcolm Gillis *et al., Taxation and Mining: Non-Fuel Minerals in Bolivia and Other Countries*, Ballinger, Cambridge, Mass., 1978. The problems of the Bolivian industry are discussed in chapters One, Two and Six.

20. Gillis, op. cit., p. 28.

21. Gillis, op. cit., p. 51.

22. Gillis, op. cit., p. 58.

23. Fox, op. cit., p. 69.

24. Knorr, op. cit., p. 275.

25. Ludwig Schätzl, *The Nigerian Tin Industry*, Nigerian Institute of Social and Economic Research, Ibadan, 1971. A more optimistic view, not borne out by events, was expressed by Peter L. Harrigan, 'Nigerian Tin Mining: Better Hopes for the Future?' in *Tin International* December 1973, Harrigan discusses the labour problems of the Nigerian industry and the impact of oil on the wages level.

26. *Tin International*, June 1980.

27. C.L. Sainsbury, 'Tin Resources of the World', *Geological Survey Bulletin*, 1301, Geological Survey, US Department of the Interior, Washington, 1969, p. 47.

28. *Notes on Tin*, August 1980, quoting the *Mining Magazine*, August 1980.

29. According to a report in *World Mining*, Chinese tin output in 1979 was running at the rate of 20,000 tonnes per annum. An editorial team from *World Mining*, after a visit to China, described the country's mining industry as 'Largely on a technical level comparable to that of Western countries 15 to 20 years ago'. See also *Tin International*, April 1981, for a summary of the paper on Tin Smelting and Refining in China by Dr Thomas S. Mackey, presented to the annual conference of the American Institute of Mining, Metallurgical and Petroleum Engineers, in February 1981.

30. Knorr, op. cit., p. 67.

31. World Bank Staff Commodity Paper No. 1, *The World Tin Economy: an Econometric Analysis*, June 1978. See also ITC, *Tin Production and Investment*, 1980.

32. World Bank, op. cit., p. 11.

33. World Bank, op. cit., p. 25.

34. The World Bank study also looked at the sensitivity of the tin price to world inflation. With the US Wholesale Index as a proxy for world inflation, the estimated coefficient indicated an inflation elasticity of about 1.5 per cent, which implied that a 1 per cent increase in the rate of inflation, thus measured, was associated with a 1.5 per cent rise in the price of tin, p. 25.

35. ITC, op. cit., p. 116.

36. ITC, op. cit., p. 120.

37. ITC, op. cit., p. 120.

4 TIN CONSUMPTION, MATERIALS SUBSTITUTION AND TIN-ECONOMIZING TECHNOLOGY

Since the late twenties, world consumption of tin has grown more slowly than the consumption of most other industrial materials. Average consumption in the years 1925–9 was 154,000 tonnes. Fifty years later it had grown by only one-third. During the early postwar years consumption was at a much lower level, averaging only 112,000 tonnes from 1945 to 1949, over 70 per cent of which was accounted for by the US and the UK. In the last 30 years consumption has certainly risen substantially, but at a much slower rate than consumption of copper, zinc, lead, aluminium and nickel, and most of the growth occurred in the fifties and sixties. Throughout the seventies, consumption has been maintained above the level of all but the best years before the war; nevertheless, the impression is inescapable that there has been a sustained lack of buoyancy in the world tin market during a long period of industrial growth and rising living standards in the industrial world.[1]

It might be assumed from the list of uses that tin would have gained more from the expansion of high-income industrial economies in the postwar world, even if the amount of tin used per unit of the total product is very small in all important applications. Looking at the prewar pattern of consumption in the US, K.E. Knorr came to the conclusion that of total consumption, 'roughly one-half is largely determined by the development of total consumer income, roughly one-quarter by supernumerary (discretionary) income, and the remaining quarter by the volume of investment in producer goods and in building construction'.[2] A recent World Bank econometric study of the world tin economy from 1955 to the mid-seventies concluded that the strongest explanatory variable of world tin consumption was industrial production. It noted, however, that 'the large number and great variety of uses for tin, other than the production of tinplate, make it difficult to capture all the economic forces which determine the demand for tin in these end-uses in econometric relationships.'[3]

From the point of view of individual countries, a high degree of industrialization and a high GNP per head do not necessarily mean a high tin consumption, unless allowance is made for the import of manufactures with a tin content. Two excellent examples of advanced industrial countries with a relatively low tin consumption are Sweden and Switzerland. The chief reason for this apparent contradiction is that both countries lack tinplate capacity, which is the main source of tin consumption in industrial countries.[4]

The question naturally arises of the reasons for the slow long-term growth of tin consumption. Compared with the rates of growth over the previous 50 years, there was a marked deceleration from the twenties to the seventies. In the last 30 years of the nineteenth century consumption rose by 50 per cent a decade. After a 30 per cent increase in the first decade of the present century, consumption slackened off in the next ten years, but increased by another 50 per cent between 1920 and 1929. Apart from one year in the thirties, the 1928–9 average was not regained until the late sixties.

It could be argued that as a very old industrial material, tin would be vulnerable in certain uses to changes brought about by the rapid growth of scientific and technical knowledge in the present century. New materials and new technologies have appeared, and some tin-using products have diminished in importance. On the other hand, the range of metal-using products has greatly increased, which might be expected to offer some compensation for the adverse effects of a changing industrial pattern. It would be a mistake, therefore, to put too much emphasis on the disadvantages of an early start.

Prices and Consumption

Compared with other base metals, tin is essentially an expensive industrial material (see Fig. 4.1); as such, it has long been vulnerable to competition. K.E. Knorr commented in his 1944 study of the tin market that the fundamental expensiveness of tin provided a constant incentive to substitution, and argued that it would be even more vulnerable in the future if there were a sustained rise in its price. In fact, according to Knorr, 'substitution would be encouraged even if the price were permanently stabilised at, say, its average for the last twenty years'.[5] That not much substitution had occurred up to his time of writing, was due, in his opinion, to the state of technology and to the expected capital costs of change, but he foresaw a different situation after the war.

While tin was much more expensive than other volume non-ferrous metals long before the Second World War, it had not always been so.

Figure 4.1: Average Annual Tin Prices (£/tonne) 1850-1980

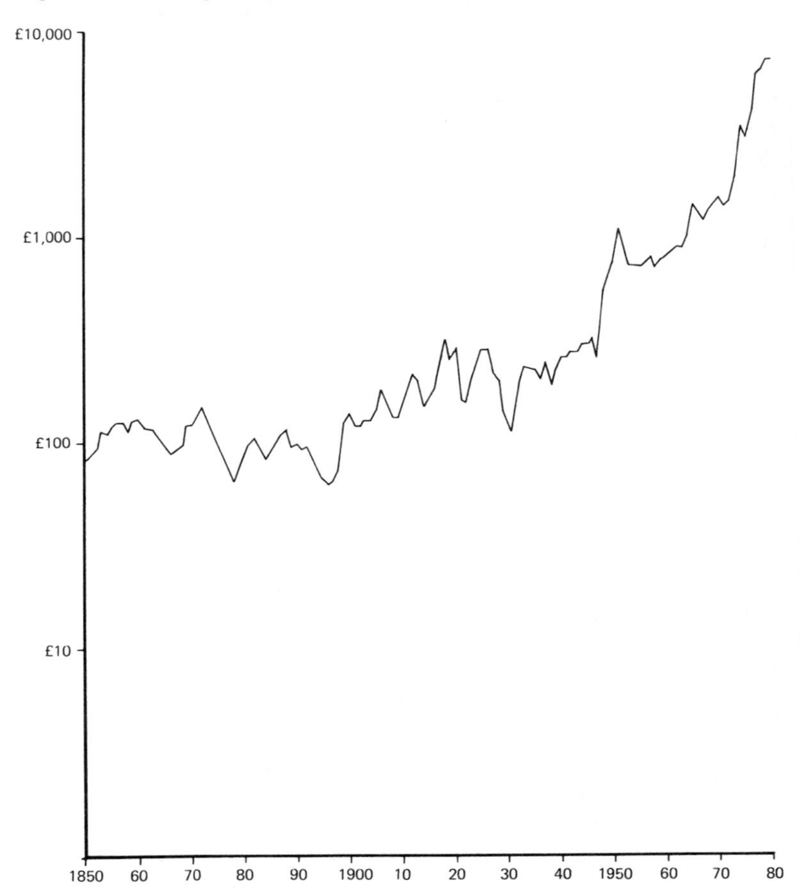

Table 4.1: Non-ferrous Metal Price Ratios, 1790–1980[a]

	Tin/copper	Tin/aluminium	Tin/lead
1790	0.8		4.5
1800	0.6		5.3
1810	0.9		5.4
1820	0.5		3.4
1830	0.8		6.0
1840	0.8		4.0
1850	0.9		4.7
1860	1.3		6.1
1870	1.8		6.8
1880	1.3		7.3
1890	1.6		8.0
1900	1.8		8.1
1910	2.5		11.9
1920	2.7	1.8	7.8
1930	2.3	1.5	7.8
1940	4.1	2.3	14.8
1950	4.2	6.5	7.0
1960	3.2	4.3	11.0
1970	3.2	11.7	11.4
1980	7.6	–	–

Note: a. Based on average annual London prices in £s per tonne.

Sources: Schmitz, C.J., *World Non-Ferrous Metal Production and Prices 1700–1976,* Cass, London, 1979. ITC, *Monthly Statistical Bulletin*.

If copper is used for comparison, the turning point in the price relationship with tin seems to have occurred around the mid 1850s. Until then, tin was generally cheaper than copper, sometimes substantially cheaper, but after the 1850s, except for 1876–8, copper was the cheaper metal (see Table 4.1). In the first decade of the present century the tin-copper price ratio fluctuated around two to one. It rose sharply in the 1920s, then fell back at the end of the decade, but to a value not much out of line with the average in the early years of the century. Throughout the thirties and forties the price ratio was generally about double the average for the first decade of the century, and particularly high in the mid-thirties. During the fifties, sixties and early seventies, the ratio was around 3.5 to one, considerably lower than in the previous two decades, but well above the values of the early part of the century. From 1975 to 1980 it was much higher than in any previous year, reaching a peak of about 9.5 to one in 1978.

A comparison with lead (Table 4.1) shows a shift in the price ratio from the late eighties. Tin had always been much dearer than lead, but towards the end of the century the price ratio edged up to eight to one compared with around five or six to one previously. In the first decade of this century it was fairly stable at about ten to one. From 1910 to 1931 it fluctuated violently, but was quite often less than eight to one. In the late forties and throughout the fifties the ratio was considerably more favourable to tin than at any time since the twenties, probably because lead was relatively scarce in these postwar years. The price ratio changed again after 1961, with tin becoming relatively much dearer, particularly in the late seventies.

Compared with the newer metals, nickel and aluminium, tin has been much more expensive in all but a few years (see Fig. 4.2). The tin–aluminium ratio has been important, because the two metals are substitutes for each other in certain uses. The price ratio moved to the disadvantage of tin in the forties, but did not change much in the fifties and up to 1963, since when there has been another shift in favour of aluminium, with typical values of around ten to one, more than double those prevailing in the fifties.

In all important uses the cost of tin is only a small proportion of the price of the final product to the consumer. This might suggest that demand for tin would not be very responsive even to a substantial increase in its price. It is not likely, for example, that a 50 percent increase in the price of tin would make much difference to the price of a £5,000 car, in which a small amount of tin may be used for radiator solder and body filler. Nor would it make much difference to the price of canned food, for which the effect on the retail price would be less than 1 per cent. The final user does not mind what is used for the radiator or the body of a car as long as it is effective. As long as the canned food satisfies all the usual requirements, the consumer is equally indifferent to the can-making material.

For these reasons Yip Yat Hoong has argued that the demand for tin is inelastic. [6] In his view, the value of the tin content of manufactures is 'virtually insignificant'. Knorr came to a similar conclusion in his study of the prewar tin market, but qualified his conclusion with the warning that demand might become more price-elastic in the future as the result of technological change, of which there were already indications in the thirties.[7] The recent World Bank study found that 'the price of tin was another variable which was statistically significant in almost all the estimated demand equations for non-tinplate end uses'.[8] But although 'statistically significant', the elasticities were low in the short run, and varied widely in the long run, so that it was difficult to draw any conclusions.[9]

There are reasons for believing that price increases can be damaging to tin, even although its cost is only a small percentage of the total cost

Figure 4.2: Tin:Copper and Tin:Aluninium Price Ratios, 1900-1980

of manufactures. A former director of the Tin Research Institute, which is concerned with the technical promotion of tin, has pointed out that the influence of price comes about through the decisions of technologically trained managers.[10] Their decisions on what material to use are taken at stages of production where the material cost is a higher percentage of cost than at the final user stage.

At intermediate stages of production the cost of the tin content is related to the cost of alternatives, including the capital cost of switching. If tin is judged to be an expensive material, there will naturally be an incentive to use less of it. However, the necessary technology may not have been developed, as was the case in the nineteenth century when Welsh tinplate producers were concerned with the high cost of tin. In the first half of last century tin is estimated to have accounted for about 18–20 per cent of the cost of tinplate.[11]

The position in the US has recently been examined for several uses of tin, as part of a wider inquiry into materials substitution.[12] The cost of tin for tinplate manufacture has been expressed as a percentage of total raw material costs when tin was US$5.25 a pound and steel US$282.88 a ton. The authors of the study argues that at an estimated 8.4 per cent of material costs there is clearly an incentive to economize in the use of tin. The study gives a good illustration of the escalation of the cost of tin on the way back from the final purchase price of canned soft drinks to the ultimate consumer. At the retail level in the US the tin cost to the consumer is indeed minute, a mere 0.5 per cent of the retail price. To the beverage producer it is 0.8 per cent of the purchase price of the empty can, to the maker of the can 2.3 per cent, and to the tinplate producer 3.6 per cent of the total cost of making the tinplate. But to the last-mentioned the decisive figure was the 8.4 per cent of the raw material cost in an industry with high capital and low labour costs.[13] This is a lower figure than the one given in a recent study of the British tinplate industry, for which tin costs are expressed as a percentage of average direct costs in tinplate manufacture. In the British case, for which the data may not be comparable with that of the US study, the tin cost was about 20 per cent of the material cost and 13 per cent of average direct cost.[14]

Economy in the use of tin has been a persistent objective of US steel producers for many years. The growth of intense competition for the huge American packaging market has forced tinplate producers, and can manufacturers, to minimize wherever technically possible the raw material input in tinplate and cans. Since tin, tonne for tonne, is the most costly material in tinplate, it is an obvious target for economy.

Moreover, steel producers are primarily interested in selling steel. They are committed to tin only to the extent that it remains the most cost-effective coating for steel plate in the metal packaging business. As their chief aim is to hold on to as large a proportion of the packaging market as possible, this means keeping down the cost of both steel and tin in tinplate. The can-makers also are actively engaged in the battle to maintain their share of the packaging market. Both they and the steel producers face a market situation which has been described in the following terms:

> Pressures from within the manufacturing chain of tinplate maker and can maker, pressures from alternative forms of packaging materials such as aluminium and plastics, pressure from alternative containers such as bottles and jars, pressures from alternative food processing systems such as freezing and dehydration, pressures from customers threatened by alternative packaging and processing developments, pressures from customers threatened by direct competition, and finally – the ultimate pressure – the need to ensure that the canned food – as a non-essential good – continues to compete for the disposable income of the consumer.[15]

If tin is not only a continually expensive material, but also a material which is tending to become relatively more expensive, it must become more vulnerable to competitive forces. Short periods of high prices, in themselves, would not necessarily be seriously damaging to a multi-purpose commodity. Other non-ferrous metals have shown a similar liability to price fluctuations, with sometimes very lengthy periods of high prices. All materials with a record of price instability tend to suffer in comparison with more stable materials.[16] A long-term upward price trend, however, is likely to have a serious effect, if there is the possibility of substitution from cheaper materials or of simply using less tin for the same purpose.

There was a period in the early and mid-twenties when a tin shortage led to an abnormally high price. After the early thirties, when tin was hit hard by the commodity slump, the price made a sharp recovery, and remained high, in the conditions of the time, for several years. From 1950 to 1963, the price rose more slowly than the prices of competing materials such as aluminium, chromium and glass, then the position was reversed during the rest of the sixties and throughout the seventies. There was clearly no hope of competing with aluminium on a price basis in some uses where markets had been lost in the forties.

Taking the postwar period as a whole, it appears that the competitive position of tin has undoubtedly been weakened. Fox reached this conclusion in his sympathetic study of the international tin agreement, and other observers such as Tilton, Smith and Schink agree. In fact, the competitive position of tin became even weaker towards the end of the seventies when its price held up much better than the prices of other non-ferrous metals in the more depressed conditions of the time. The sharp increase in the tin–copper price ratio showed the marked contrast between the experience of the two metals.

As the long-run trend in the price of tin has made it relatively more expensive, technological innovations have eroded part of its market. The first uses to be affected have been those where tin has made the least specific contribution to the product. Other uses have been affected only after extensive research into the applicability and cost-effectiveness of substitutes, thus delaying the switch. Tin has certainly not been the only non-ferrous metal to lose ground to innovation the last few decades. Copper, lead, zinc, and even aluminiun have all been hit by materials substitution in various forms and for various reasons. What is most striking about tin, however, is the powerful impact of substitution over a long period, which has given it such an unusually low growth rate.

The process of materials substitution works in several ways, usefully classified in the Tilton study of the tin market in the US. Tilton lists three methods, with several sub-divisions.[17] Firstly, there can be simply material for material substitution, where one material is substituted directly for another. Aluminium car radiators might replace the conventional copper–brass radiators. Secondly, there is functional substitution, defined by Tilton as the replacement of a product by an entirely different way of performing its function. An example quoted by Tilton is the development of frozen food packages at the expense of metal food cans. Thirdly, there is material-conserving substitution, which may take three forms: (1) technological progressive substitution by which is meant a technological improvement which allows a product to be made with less material; (2) other-factors-for-material substitution, which reduces materials consumption by increasing the use of capital, energy or other inputs in the production process; (3) quality substitution, which reduces material requirements in producing an article by a reduction in its quality or performance characteristics. Examples given by Tilton for each of these sub-divisions are less tin in beer and soft-drink tinplate cans, automatic production of printed circuit boards using more tin-containing solder than is necessary with hand-soldering, and the use of thin glass wartime bottles with high breakage rates.

Tinplate and Tin Consumption

In its principal use, tinplate, tin competes with glass, aluminium, chromium, plastics, paper and board, and with frozen foods. In developing countries particularly, tinplate food cans compete with unpackaged fresh food.

The effects of tin-economizing technology have been shown most dramatically in tinplate, which has been the chief use of tin for the whole of this century. There have been several reasons: partly the influence of the price of tin; partly government pressure on users, especially in the US at certain times; partly other non-price factors. In principle, only the first is measurable, and in practice difficult to assess with precision. However, the Wharton tin model has apparently found the impact of higher tin prices on the tin content of tinplate to be strong enough to detect econometrically. The model gave the long-run price elasticity of tin metal used per ton of tinplate as –0.27 in the US and –0.21 in Western Europe. In other words, a 1 per cent rise in the price of tin would lead in the long run to a 0.27 per cent or a 0.21 per cent fall in the amount of tin used. The same model finds that the effect of tin prices on the use of tin for solder, the second main use of tin, is statistically insignificant'.

Some experts believe that the development of electrolytic tinplate, the major tin-economizing innovation, has on balance helped tin consumption. According to a former director of the Tin Research Institute:

> It is worth reminding ourselves that the change from hot-dipped to electrolytic tinplate, viewed at one time as something of an alarming development for tin . . . has kept at bay more serious encroachments by aluminium and tin-free steel (TFS). It is fair to say that if the only tinplate available today were the hot-dipped variety, we should be using very little tin for tinplate because economic competition would have been impossible, and alternatives would have swept the board.[18]

Research into the electrolytic process of tinplate manufacture, the deposition of tin on steel plate by electrolysis as a means of cutting tin consumption, began in Germany even before the First World War.

However, it was not until the thirties that a German scientist took out the first patent for producing a bright electrolytic coating on steel. The motivating force in Germany was not so much the price of tin as the conservation of foreign currency and a greater degree of self-sufficiency. During the thirties and forties, however, new investment by German industry in what was a much more capital-intensive method for producing tinplate for canned food did not command a high priority. Artificial rubber and synthetic petrol were much more important in prewar and wartime Germany. Postwar West Germany did not begin the production of electrolytic tinplate until the fifties. Tin Council statistics do not record German production until 1958, after which progress was rapid.

Large-scale experiments with the electrolytic process began in the US in 1936 and went on until the war. Steel producers realized before the war that eventually hot-dipped tinplate would be replaced by electrolytic tinplate, but for several reasons they delayed the creation of electrolytic capacity until the war with Japan. According to Knorr, 'under normal circumstances the large-scale application of electro-deposition to the manufacture of tinplate would undoubtedly have developed but slowly'.[19] He lists the reasons for delaying the use of electrolytic tinplate in food packaging: manufacturers were reluctant in the conditions of the thirties to invest heavily in new plant and make much of their existing plant redundant; although the basic manufacturing processes were already well known, they were still in the experimental stage and needed improvements; it was necessary to carry out tests on the various thicknesses of tin coating for different varieties of canned goods; and crucially, 'no reliable comparative cost analyses of hot-dipped and electrolytic tinplate manufacture were available'.[20]

The war with Japan proved the catalyst for the start of the switch to the new tin-economizing process, just as it was to do with other uses of tin. Within a year of the federal government directive in December 1941, requiring American steel producers to cut their tin consumption, the first electrolytic line had been built, more capacity being installed in the later war years, so that by 1950, 59 per cent of US tinplate output was electrolytic.[21] The process of change was gradual; even in 1954, 12 years after the installation of the electrolytic line, the US still produced about one million tonnes of hot-dipped tinplate. By 1957, however, the share of hot-dipped tinplate in the US had dwindled to 10 per cent and in the next ten years to less than 1 per cent. Thus it took about a quarter of a century to complete this major technological revolution in the pioneering country.

Other countries did not begin electrolytic production until well after the war, but its introduction was associated with a big increase in the demand for tinplate in the food packaging industry of many countries. One developing country, Brazil, starting from scratch as a tinplate producer with hot-dipped capacity in 1948, reached the 85 per cent electrolytic share as early as 1960. By 1970 all industrial tinplate producing countries had either reached or exceeded the 90 per cent level. Only Japan and India still had a substantial amount of hot-dipped capacity. By the end of 1977, hot-dipped capacity, as recorded by the Tin Council, was reduced to only 269,000 tonnes, of which 121,000 tonnes capacity was in India. The sole British producer, the BSC Tinplate Group, stopped producing hot-dipped tinplate at the end of 1976, but small amounts of defective tinplate are repaired by hot-dipping at another company's plant. World production of tinplate in 1979 had risen to 14,286,000 tonnes (see Table 4.2) compared with 2,815,000 tonnes in 1938, and the number of countries with some tinplate capacity had risen to 38 compared with only 10 in 1938.

The switch from hot-dipped to electrolytic tinplate has been the most revolutionary change in the market for tin. It also had a major effect on the prospects for steel in the canning of food and beverages. It speeded up the production of tinplate enormously. The fact that the working unit could now be a continuous strip instead of a batch of steel sheets made it possible to streamline the plant, and eventually to introduce remote control by press button, so the amount of labour required fell dramatically. Labour productivity rose seven-fold between 1939 and 1960 in the UK. In 1939 25,000 workers were needed to produce 929,000 tonnes of tinplate. By 1960 4,300 workers produced 1,191,000 tonnes.

Another important advantage of electrolytic tinplate was that much greater variety of tin coating on the steel base became possible. Electrolytic deposition of tin allowed the controlled uniform application of thinner coatings than had been possible with hot-dipping. It also became possible to control independently the coating applied to the two surfaces. As Dr B.T.K. Barry has pointed out: 'This differentially-coated tinplate enables the user to select separately the coating best suited to the contents of the container and to the atmosphere. Thus the flexibility of the product has been greatly increased for a market which is highly diversified in its requirements for metal containers. Tinplate has become a veritable family of materials'.[22]

For these reasons tin has been able to benefit from an increased

Table 4.2: World Production of Tinplate and Tin Used, 1948-1979[a] (000 tonnes)

	Tinplate production	Tin used	Tin percentage (by weight)
1948	4,460	49.9	1.10
1950	5,646	60.2	1.00
1954	6,332	59.9	0.92
1956	7,457	66.7	0.89
1958	7,489	61.0	0.81
1960	9,164	72.1	0.79
1962	9,145	68.9	0.75
1964	9,806	74.2	0.75
1966	10,457	74.6	0.70
1968	12,301	80.3	0.65
1970	12,967	80.0	0.62
1972	12,577	76.9	0.60
1974	14,398	86.5	0.60
1976	13,513	78.6	0.58
1978	13,718	75.0	0.55
1979	14,286	75.3	0.53

Note: a. As noted in the text, the measurement of tin used by weight instead of by area of tinplate will underestimate the number of containers which can be made from a given quantity of tinplate, since tinplate grades have become thinner in the last 30 years. Some tin mill products other than tinplate are included.

Source: ITC, *Monthly Statistical Bulletin*, various issues; ITC, *Statistics of Tin*, 1945-70, and *Tin Statistics*, 1969-79.

demand for tinplate by the packaging industry, while at the same time its consumption per tonne of tinplate has greatly fallen. For an average tinplate production of 3.5 million tonnes in the 1935-9 period, tin consumption was 59,000 tonnes. An output of 14,221,000 tonnes in 1979 needed 75,300 tonnes of tin. As far back as 1950 it was possible to produce 2 million tonnes of tinplate more than in the thirties with about 2 per cent more tin. The tin coating by weight has fallen from 1.5 per cent to an average of less than 0.55 per cent. Measuring the reduction by weight is less satisfactory than measurement by area of tinplate covered by the tin coating. It would be desirable to express output of tinplate by area, since the production in the thickness of the steel base has been substantial.[23] A given quantity of tin, therefore, covers a much larger area of tinplate now than it would have done 20 or 30 years ago. It follows that the output of tin cans which can be coated with tin, especially if differential coating is used, is much larger than it used to be. Unfortunately, the figures available for world tinplate production in the Tin Council's statistical series are only in terms of tonnage.

Figure 4.3: Tinplate Production and Tin used in Tinplate, 1945–1979

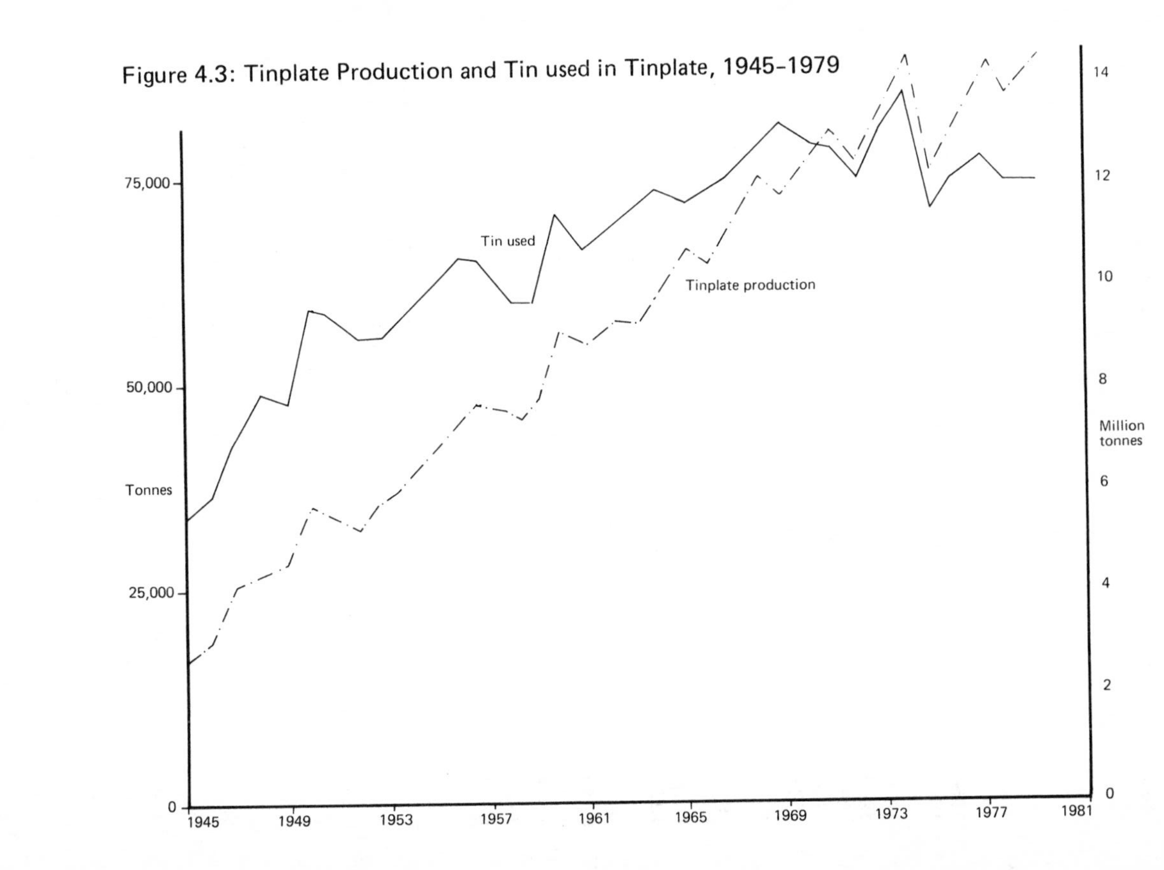

Table 4.3: Tin Used in Tinplate by Country, 1977

	Tinplate[a] production (000 tonnes)	Tin used (tonnes)	Tin percentage (by weight)
Brazil	461.7	1,785	0.39
USA	4,228.3	18,539	0.44
Belgium	362.6	1,724	0.48
Canada	533.3	2,764	0.48
UK	1,176.1	6,372	0.54
Netherlands	460.5	2,629	0.57
France	915.5	5,279	0.58
German F.R.	851.5	5,114	0.60
Italy	399.9	2,414	0.60
USSR	612.0	4,200	0.69
Japan	1,866.0	12,988	0.70
Spain	309.4	2,300	0.74

Note: a. Includes some tin mill products other than tinplate.

Source: ITC, *Tin Statistics*, 1969–79.

During the seventies, as Table 4.2 shows, the amount of tin used in tinplate has continued to fall. The conversion from hot-dipped to electrolytic tinplate was not a once-for-all switch. That it is a continuous process of change is shown particularly by US experience, where a 25 per cent fall in tinplate production between 1969–70 and 1978–9 was accompanied by a 30 per cent fall in tin consumption. Although most tinplate is now electrolytic, the average amount of tin used per tonne of tinplate varies substantially between countries. The 1977 figures, shown in Table 4.3, are lowest for Brazil and the US, highest for Japan and Spain. To a large extent such variations between countries are the result of differences in the canned products, some of which may be more acidic than others and require, therefore, a more protective tin coating on the inside of the can.

Competition with Aluminium and TFS

So far, the fortunes of tin and steel have been linked with the competition between tinplate and aluminium. US aluminium producers set their sights many years ago on the large potential market for canned beverages, the popularity of which owed much in the US to the preference of ex-servicemen for throw-away beer cans. According to Frederick R. Demler, the role of ex-servicemen is considered to be 'one of the most important factors underlying the growth of one-way containers in a returnable bottle orientated society'.[24] The great expansion of aluminium capacity after the war and the continued fall in the real cost of aluminium, which benefited from the cheapness of energy until the seventies, induced aluminium producers to compete for a slice of the can market. Their success in the US forced steel producers to spend still more on research into cheaper and better tinplate. By reducing the thickness of the steel base, they cut the cost of both steel and transport. A thinner tinplate also made for easier opening of cans, a minor but by no means insignificant contribution to the growing popularity of canned goods. The aluminium producers, for their part, developed the easy-open end and the ring-pull end for beverage cans.

Tinplate regained some competitive power during the seventies from the sharp increase in the price of energy. Estimates by B.T.K. Barry, shown in Table 4.4, suggest that, compared with aluminium and glass, tinplate needs less energy, measured in terms of BTUs per container, or in terms of the amount of energy consumed producing the raw material for one tonne of containers. According to US data, the total cost of tinplate two-piece cans is less than that of aluminium cans for alcoholic and non-alcoholic beverages (see Table 4.5). However, it is pointed out by Demler that, while the cost of an aluminium two-piece beverage container is about 12 per cent greater than the cost of a two-piece tinplate container, production costs are not the only costs influencing the decision on which material to use.[25] Transport costs are less for aluminium cans, which weigh much less per thousand cans than tinplate cans. There is also a recycling advantage for aluminium since recycling presents fewer problems for aluminium, which makes it easier to comply with environmental restrictions.

Another threat to tin has come from the efforts of steel producers to cut raw material costs by eliminating the tin coating on tinplate and replacing it with a substitute coating. Prewar research in the US had led as early as 1934 to the production of cans without a tin coating. To prevent external and internal corrosion special organic coatings were applied to blackplate, the basic steel product made in the tinplate mills. During the war large quantities of the 'plain' steel cans were produced for a variety of uses in the US, the UK and Germany. As a wartime expedient this alternative to tinplate was successful, but it was not commercially worthwhile under normal peacetime conditions. The American Can Company stated that lacquered blackplate cans were not as efficient as tinned cans, but it continued research after the war when there were no longer restrictions on the use of tin. The attitude of American users was shown in the American Can Company

Table 4.4: Energy Consumption in Container Production

	Energy consumed producing raw material for 1 tonne of containers (Gigajoules)	No. of containers per tonne	Energy consumed per container (BTU)
Tinplate can	49	16,500	2,850
Aluminium can	395	44,500	8,210
Bi-metallic can	77	18,400	3,990
Glass bottle (returnable)	54.8	2,000	26,100
Glass bottle (non-returnable)	54.8	4,000	13,050

Source: Dr B.T.K. Barry, 'Tinplate 1978 — an international perspective', paper read to an Australian Tinplate Conference, Sydney, 1978, reported in *Tin International*, December 1978.

Table 4.5: US Container Costs (aluminium and steel two-piece can bodies 209/211 x 413[a] – US$ per 1,000)

	Aluminium (33 ≠ ≠)[b]		Tinplate (118 ≠ ≠)[c]	
Cost element	Beer	Soft drinks	Beer .25[d]	Soft drinks .25[d]
Metal				
Starting	31.75	31.75	22.98	22.98
Scrap credit	3.88	3.88	.53	.53
Net	27.87	27.87	22.45	22.45
Cleaning and treatment	.30	.30	.10	.10
Interior coating	.65	.90	.975	1.19
Tooling costs	.15	.15	.35	.35
Labour cost	4.93	4.93	5.11	5.11
Overhead cost	2.33	2.33	2.41	2.41
Depreciation	3.58	3.58	4.20	4.20
Total cost	39.81	40.06	35.60	35.81

Notes: a. Refers to volume of beverage can (12 oz). b. Weight of aluminium sheet per box (33 lb). c. Weight of steel in tinplate per base box (118 lb). d. Weight of tin per base box (.25 lb).

Source: Frederick R. Demler, 'Beverage Containers', in Tilton (ed.), *Material Substitution: the Experience of Tin-Using Industries*, Penn. State Univ. Park, Penn., 1980, p. 40.

announcement that it was 'engaged in a battle of research to free itself from dependence upon foreign tin supplies.'[26]

During the fifties and sixties the Japanese also were engaged in the search for a commercially successful alternative to the tin coating. Eventually, in the late sixties, chromium-coated steel emerged as the only serious challenger. Chromium was much cheaper than tin, and less was needed for the same purpose. In 1967 it was about one half the price of tin.[27] In contrast to most other metal packaging innovations since the last war, the invention of chromium-coated steel for cans was not an American achievement, both the commercially successful processes for coating steel being developed in Japan. Chromium-coated steel, or tin-free steel (TFS) as it is rather unfairly called, hit the trade headlines in the late sixties. Before the end of the sixties, practically all the leading tinplate producers in the world had taken out TFS licences.[28] Forecasts were made at the time of a drastic fall in the consumption of tin by the steel industry, yet by 1978 it was possible for one impartial authority to wrote: 'There can be few innovations which have been announced with such fanfare and then been so widely licenced, that have had so little ultimate impact'.[29]

Tin producers were generally sceptical about the claims made for TFS. Some market loss was expected, as there has been to aluminium and plastics, but the total market was believed to be expanding, and had long been shared with glass and other packaging materials. TFS, moreover, had certain disadvantages, notably the fact that it could not be economically soldered. This disadvantage was eventually disposed of by the introduction of the welded or cemented TFS can. It was also necessary to spend more on lacquers.

The tin producers' confidence that the steel industry would not desert tin seems to have been borne out by the experience of TFS and tinplate in the seventies. Although precise information is not readily available, it is apparent from the US, at least, that the growth of TFS output did not maintain its early momentum. After reaching 592,000 tonnes in 1969 and more than doubling by 1973, TFS has not further increased its share of the market. Production in 1977-8 was about the same as in 1971-2. TFS seems to have made little impression on the UK market. Output in 1979 was reported to be only about 62,000 tonnes compared with a tinplate output of over one million tonnes.[30] According to a recent study of the UK packaging market, tinplate still had about 96 per cent of the market for open top and general line containers. Capacity has certainly been created for the production of TFS in countries other than the US, UK and Japan. Thomas B. Barry lists TFS production lines in Brazil, France, West Germany, the Philippines and Venezuela. Some countries have dual TFS/tinplate lines, enabling switching from one to the other according to relative coating and other costs. Unfortunately, it is difficult to assess the total penetration of the market by the various forms of TFS.

Tinplate production continued to expand in the 70s, but with fluctuations which reflected the more disturbed economic conditions of the decade. World consumption of tinplate was about 12 per cent greater in 1979 than in 1970, with a 5 per cent smaller tin consumption. Clearly, forecasts in the last sixties that tinplate would have lost heavily by the end of the seventies were completely inaccurate.

That tin has held on to most of its share of the packaging market is due to a number of factors. Switching from tinplate to TFS involves capital expenditure for steel producers and can makers, which must be set against the fact that the material cost advantage of TFS or tinplate is believed to be quite small, if extra lacquering costs of the cans are included. Variations in the relative costs of tin and chrome affect the choice of coating. Over much of the can-making and can-using market tinplate has a strong historical position. Steel producers have continued to improve tinplate, which is a high-value, profitable part of the steel industry. Tinplate technology has not stood still. According to Barry, steel producers are looking for a substitute for TFS itself, which will use less tin than the normal tinplate and can be used on the conventional tinplate lines.[31] This suggests that steel producers are far from simply fighting a rearguard action in defence of tinplate.

At the end of the seventies, therefore, it seemed that chromium was only a marginal threat to tin beyond the share of the metal container market which it had already won in the US. But the competitive position between aluminium, tinplate, TFS and other packaging materials remained as fluid as ever. Looking only at metal cans, one user has commented: 'None of us is really sure which metal is best. The comparative economics between aluminium and steel are often misleading or at least slanted to prove a particular point. This is not hard to do, as it is difficult to get a true comparison when comparing these metals'.[32]

Substitution and Other Uses of Tin

Competition between tin and other materials has been active in other uses, but without the same large-scale effect as in tinplate. Pressure from government played an important part at times in stimulating materials substitution and tin-economizing technology. Doubts about the long-run security of supply seem to have influenced some major users. Technological change has tended to be biased against tin, even if the process has not been entirely one way. Relative prices have also had some adverse effect on tin's competitive position.

The first indications of competition from aluminium appeared in the thirties. It is unlikely that tin could have held off the challenge from aluminium in some uses even if it had been cheaper. There was a powerful underlying trend in favour of aluminium, which was becoming cheaper in relation to a number of materials, not merely tin. Copper and steel were also subject to increasingly effective competition from aluminium in a number of important markets.

During the thirties, tin foils, once used by the tobacco, chocolate and cheese industries for packaging, were increasingly replaced by aluminium, lead and zinc foils, as well as by waxed papers and cellophane. Collapsible tubes, of which pure tin was the chief component, became a classic case of wartime substitution by aluminium in the US. American data on collapsible tubes and foil show that consumption of tin fell from about 8,000 tonnes in 1928 to between 5,000 and 6,000 tonnes in the late thirties. In 1943 it was down to a mere 500 tonnes, clearly a major response to the government's directive on tin economy, the reduction in tin consumption apparently having occurred in a matter of months. Research had already been carried out into the use of aluminium for collapsible tubes, but how long the switch would have been delayed without government pressure is naturally uncertain. Possibly the need for capital expenditure had prevented a switch before the war, but once the expenditure had been incurred and aluminium was clearly shown to be the cheaper and equally cost-effective material, there was no return to tin on any significant scale after the war.

Data on tin consumption for foils is not available for the UK in the thirties and forties up to 1946. It seems, however, that consumption in collapsible tubes fell sharply in 1949-51, since when it has never been more than half the prewar rate. Tin Council statistics show that consumption has been stable at a low level since the early fifties. That some tin is still widely used in collapsible tubes reflects its special properties for certain medicinal preparations, but the total amount used is now only a small percentage of total tin consumption. Fortunately for tin producers, it was never more than 8 per cent of consumption in the two leading prewar consuming countries.

A much more important non-tinplate use of tin has been in solder and other alloys. For tin to remain a volume material, if on a much smaller scale than other important non-ferrous metals, it has been necessary for tin to hold on to the market for these uses, together with the tinplate market.

There has been a substantial loss to tin in the solder market. This is evident from the way in which tin consumption has failed to keep pace with the great expansion of output of metal-using products over the last 30 years. In the US, for example, the amount of tin reported in solder was no higher in the sixties than in the twenties. Consumption in 1973, the peak postwar year, was only slightly higher than in 1950. Tin Council estimates of consumption in ten industrial countries and in Brazil show no significant signs of growth between 1965 and 1978.

The use of tin for solder reflects changes in both the tin content of solder and the use of solder itself. A detailed study of tin and the solder market in the US by Patrick D. Canavan lists four determinants: the number of end products (cans, radiators, automobile bodies, etc.) produced; the proportions of soldered products in that total; the quantity of solder used per unit of soldered end product; the average tin content (as a percentage of weight) of the solder used for the end product.[33] The first determinant is treated as exogenous by Canavan and, therefore, little attempt is made to explain changes in it.

Canavan has found that up to the war not much change seemed to have occurred in either the tin content or the amount of solder per unit of output. Once again there was a sudden impact from the directive on tin consumption by the US War Production Board, which specified that the tin content of all solders must be limited to 30 per cent, a maximum which was further lowered to 21 per cent by early 1944.

Months before Pearl Harbor, strategies for tin conservation had been discussed, and solder had been looked upon as a major target for economy. In spite of some transitional difficulties the switch to lower tin solders was successful. Canavan's study of the US market shows the effects on a selected number of uses for solder, covering a high proportion of all solder used in the US. His case studies included metal cans for fruit, vegtables, alcoholic beverages, evaporated milk, motor vehicle radiators, automobile body fillers, plumbing and aerosols. In all the products manufactured during the forties there was a sudden wartime cut in tin consumption per unit of output, not subsequently reversed.

Substitutes had been considered for high tin solder long before the war, but there had been doubts about their practicality, consistent with preserving the necessary physical properties of solder. The costs of a breakdown in a finished product would have been far greater than any economy in the use of tin. Since the wartime compulsory cuts in the tin content had no ill effects on the efficiency of solder, it looks as if the prewar traditional practices used more tin than was strictly necessary. The cuts were drastic. It is not surprising, therefore, that the Paley Commission in 1952 was sceptical about even more drastic cuts. It commented: 'tin-lead solders are so widely and easily used that substituting other materials is probably more difficult than for any other general application of tin'.[34]

For many years after the war this conclusion by the Paley Commission seemed justified. Canavan found that 'the aggregate tin content of solder has remained quite static, varying between 21 per cent and 26 per cent tin since 1950, with only the years 1973 and 1974 falling somewhat outside this range'.[35] Canavan also points out that the unchanged composition of solder seemed to be unaffected by the fact that between 1950 and 1978 the ratio of the tin price to the lead price rose from 7.5 to 18.7. It seemed to follow from this deterioration in the competitiveness of tin that 'with such price incentives to reduce the use of tin, the constancy of composition suggests that the substitution of other materials for this is indeed infeasible'.[36]

Looking at the use of solder in detail, however, Canavan found that this conclusion was unjustified. It was the result of aggregating the various uses of solder and their tin content. An examination of the trend in the tin content of particular uses revealed sharp changes, some solders with a high tin content at one time now using less, but with the average being maintained by the great expansion of the electronics industry, which used a high 60–63 per cent tin solder almost exclusively, at least in the US. Probing further, Canavan found that even before 1950 there was 'much less support for the premise of insubstitutability even in the aggregate data'.[37]

The growth of the electronics industry has clearly been of vital importance for solder consumption and the amount of tin used in solder. It has helped to offset the decline in traditional uses of solder, such as plumbing, where the use of large quantities of expensive tin metal to form a single joint, as in the old-fashioned plumbing system, or the use of tin alloys to fill up space as in a fillet, has virtually disappeared, or at least been replaced by alloys with a much lower tin content. Nevertheless, in spite of the electronics boom, there has been a continued decline in the US in the amount of solder and its tin content per unit of industrial output.

Although the enormous expansion of the electronics industry certainly helped tin, the development of technology in the industry by no means increased tin consumption in all cases. Expansion of output meant that the number of soldered joints increased, but the trend towards miniaturization of sophisticated electronic products worked to reduce demand for tin. The first computer in the US, ENIVAC, needed about 30 lbs of tin; much smaller modern equipment uses only a minute amount. Summing up the position in the US, Canavan states:

> The technological development of automated printed circuit board soldering for electronic products has profoundly affected solder and tin in solder usage. With the initial adoption of this new production technology, solder consumption increased five- to ten-fold over hand-soldered consumption levels. Subsequent miniaturization of electronic components and the printed circuit board of typical consumer electronic products has reduced the intensity of solder use to one-fifth of these levels. [38]

Of the other alloys, bronze and brass seem to have become less important, according to Tin Council data covering the main consuming countries since 1965. Table 1.3 in Chapter 1 shows a decline from over 22,000 tonnes, or 14.4 per cent, in 1965, to 9,500 tonnes, or 6.7 per cent, in 1979. Like solder, bronze and brass are recycled, and there is, therefore, a secondary tin content which, according to US data, is always greater than the consumption of primary tin in bronze and brass. American experience suggests that the use of tin in bronze and brass has fallen chiefly because there is less secondary consumption. From 1965 to 1979 consumption of secondary tin in bronze and brass fell from over 12,000 tonnes to around 6,000 tonnes, while consumption of primary tin fell from 4,500 tonnes to less than 3,000 tonnes. In the fifties, US secondary tin consumption in bronze and brass averaged 14,500 tonnes. Tin Council data, going back to the twenties, shows that US consumption of primary tin in bronze and brass has usually been within the 5,000 to 3,000 tonne range. West European experience also shows severe competition for brass, with subsequent repercussions on tin consumption. Consumption of brass in the motor industry has lagged well behind the output of vehicles. In other consuming industries demand for brass has been affected by miniaturization and substitution. The household appliance industry has cut brass consumption, as has the mechanical engineering industry, replacing brass with stainless steel or plastics.[39]

After electronics, chemical compounds have been the main new source of demand for tin. US consumption has risen from a few hundred tonnes in the early fifties to around 5,000 tonnes. UK consumption has fluctuated since 1965 with no apparent trend, but at least has not fallen like the other main categories of use in line with the generally weak industrial performance of the economy. Consumption in West Germany and France appears to have doubled between 1973 and 1978. There is an insufficient breakdown in the Tin Council statistics of Japanese consumption to identify the chemical uses of tin, but assuming that it is the main component of 'other uses', there was at least a doubling of consumption between 1969 and 1979.

Notes

1. Copper consumption doubled between 1920 and 1929, and increased fourfold between 1929 and 1976. Zinc consumption doubled between 1920 and 1929, and increased nearly three and a half times between 1929 and 1976. Lead consumption doubled between 1920 and 1929, and again between 1920 and 1976. Nickel consumption increased nearly thirteen times between 1930 and 1976. Average annual rates of growth were as follows for five leading metals (percentages):

Copper	3.9	(1955–77)
Nickel	6.5	(1950–74)
Aluminium	7.3	(1960–76)
Iron ore	3.6	(1960–76)
Tin	1.0	(1955–76)

2. K.E. Knorr, *Tin under Control*, Food Research Institute, Stanford Univ., California, 1945, p. 45.

3. World Bank Staff Commodity Paper No. 1, *The World Tin Economy: An Econometric Analysis*, The World Bank, Economic Analysis and Projections Department, June 1978, p. 17.

4. The ITC points out that its estimates of tin consumption by country 'refer to the consumption of tin metal by the manufacturing industries of the country specified and do not necessarily represent the final domestic consumption of tin, since a proportion of the goods manufactured may be exported for final consumption elsewhere, or tin-containing goods of foreign manufacture may be imported for home consumption'.

5. Knorr, op. cit., p. 44.

6. Yip Yat Hoong, *The Development of the Tin Mining Industry of Malaya*, University of Malaya Press, Kuala and Singapore, 1969, p. 10.

7. Knorr, op. cit., p. 44.

8. World Bank, op. cit., p. 17.

9. The price elasticity of demand for tin has been analysed on both a regional and an end-use basis by the World Bank study. The study's findings on short-term and long-term country or regional demand elasticities are as follows:

Tin Demand Elasticities

Country or area	Price elasticity of demand	
	Short-term	Long-term
US		
Tinplate uses	0.24	n.a.
Non-tinplate uses	0.13	n.a.
Western Europe		
Tinplate uses	0.11	n.a.
Non-tinplate uses	0.30	0.41
Japan		
Tinplate uses	0.18	n.a.
Non-tinplate uses	0.49	n.a.
Other developed countries	0.37	1.60
Developing countries	0.11	0.15

Source: World Bank Staff Commodity Paper No. 1, *The World Tin Economy: An Econometric Analysis*, Economic Analysis and Projections Department, World Bank, Washington, DC, June 1978.

10. See the discussion by E.S. Hedges, *Tin in Social and Economic History*, Edward Arnold, London, 1964, Ch. 3.

11. W.E. Minchinton, *The British Tinplate Industry*, Clarendon Press, Cambridge, p. 56.

12. John E. Tilton (ed.), *Material Substitution: the Experience of Tin-Using Industries*, Pennsylvania State University, University Park, Penn., 1980, Ch. 4.

13. Frederick R. Demler, 'Beverage Containers', in Tilton, op. cit., p. 33.

14. ITC, *United Kingdom – Tin in Tinplate*, a report by U. Yahya, ITC Tin Industry Officer, April 1981, p. 24.

15. *Tin International*, December 1979.

16. According to the British Steel Corporation, the instability of the price of tin is 'a cause for concern because other elements of cost, while sometimes increasing at a greater rate, are more predictable and can be allowed for in budgeting and pricing'. See U. Yahya, ITC Tin Industry Officer, United Kingdom, *Tin in Tinplate*, ITC, London, April 1981, p. 24.

17. See the introductory chapter by Tilton, op. cit., pp. 3–4; also his general summing-up on substitution, pp. 252–70; and his preliminary review of the study, 'Understanding Material Substitution', *Earth and Mineral Sciences*, vol. 48, No. 7, April 1979.

18. W.E. Hoare, 'Trends in Tin Consumption: Some Technological Observations', a paper to the World Conference on Tin, Kuala Lumpur, 1972, Vol. 4 of the Conference Proceedings, *Marketing and Consumption*, p. 16.

19. Knorr, op. cit., p. 19.

20. Knorr, op. cit., p. 19.

21. The speed and extent of wartime economizing on tin consumption in the US, powerfully influenced by government pressure, are described by Knorr, Ch. 13.

22. See B.T.K. Barry, 'The World Tinplate Industry', a paper to the World Conference on Tin Consumption, London, 1972; also the summary of the Sydney Conference on tinplate in *Tin International*, December 1978.

23. A. La Spada, Statistician to the ITC, made this point strongly in a paper to the International Tinplate Conference, London, October 1980. He pointed out that the presentation of tinplate statistics in terms of tonnage rather than of area, placed inevitable limitations on the Council's statistical picture of the industry, as did the broad groupings of steel-based can stock under the heading of 'tinmill products'. At the same conference he forecast tinplate production of 16 million tonnes within the present decade, an increase of about 14 per cent over the 1978–9 average.

24. Demler, in Tilton, op. cit., p. 58.

25. Demler, p. 39.

26. Demler, p. 57.

27. Demler, p. 30. According to Demler, the price of chrome in 1977 was 2.65 dollars per pound, compared with 6.95 dollars for tin. The Schmitz statistical series gives an unchanged UK price range of £734–785 a tonne from 1962 to 1968. Over the same period tin rose from £896.5 to £1,353.3 a tonne. The UK price of chrome more than doubled between 1968 and 1976, but was still less than the price of tin in 1976. See Christopher J. Schmitz, *World Non-Ferrous Metal Production and Prices 1700–1976* Cass, London, 1979.

28. It is somewhat ironic that, in using chrome instead of tin, the US was using one of the two chief metals in which it is largely deficient (except for stockpiles, which are exhaustible). The main current producers of chrome are the Soviet Union and South Africa.

29. Thomas B. Barry, 'The Competitive Challenge to Tin', *Tin International*, November 1978, p. 427.

30. See the report to the Tin Council by Umar Yahya, ITC Tin Industry

Officer, *The UK-Tin in Tinplate*, London, 1981.

31. Barry, op. cit., p. 427.

32. Reported in *Tin International*, December 1978. According to the speaker quoted, 'one of the largest users of cans has stated that they never again want to be dependent on only one metal'.

33. Patrick Canavan, in Tilton, op. cit., p. 89.

34. Canavan, op. cit., p. 76, quoting the US Paley Report, 1952.

35. Canavan, op. cit., p. 76.

36. Canavan, op. cit., p. 76.

37. Canavan, op. cit., p. 76.

38. Canavan, op. cit., p. 176.

39. See ITC, *Notes on Tin*, February 1981, quoting the *Metal Bulletin*, January 1981.

5 CONSUMPTION TRENDS AND FLUCTUATIONS IN THE LEADING COUNTRIES

The United Kingdom

The Welsh Tinplate Industry and the US Market

For the most of the nineteenth century the UK was the largest tin consumer. During the third quarter of the century the tinplate industry of south Wales became an increasingly important consumer, and expansion continued up to 1981. At that time, most UK output of tinplate was destined for foreign buyers, over two-thirds being exported, the US alone taking about 55 per cent. Assuming an average tin coating amounting to 1.5 per cent by weight with the hot-dipped process, the Welsh industry probably used about 4,600 tonnes of tin in 1880. Total UK consumption is uncertain, but an estimate can be made by considering domestic tin output, imports and re-exports of tin metal. Imports of tin concentrates for smelting were negligible until 1883. As tin production in Cornwall in 1880 was 8,900 tonnes of metal, and imports of metal were 19,800 tonnes, the re-exports of tin blocks of 8,800 tonnes leave an apparent domestic tin consumption of 20,000 tonnes in this year.

Although the use of tinplate for food cans was growing in the second half of last century, a much higher percentage than now was used for other purposes such as domestic utensils, dairy and sugar refining equipment, baths, tea urns and metal trunks. The expansion of UK tinplate production for these and other purposes depended heavily on the growth of the US market. An indication of the growth of the US food market was the rise in the number of fruit and vegetable canners in the US from a mere 97 in 1871 to 411 in 1880 and 886 in 1890. This meant new demand for Welsh tinplate.

The growth of the US tinplate market for the Welsh industry came to an abrupt halt with the McKinley Act of 1890, which put a substantial tariff on tinplate imports.[1] As the Act involved a duty equivalent to 10 shillings on a box of tinplate, then selling for 14s 4d, the American importer now faced a 70 per cent increase in the cost of imported tinplate. UK exports to the US fell from 325,000 tonnes in 1891 (out of a total output of 586,000 tonnes) to only 60,000 tonnes in 1900 and 21,500 tonnes in 1913. Fortunately for the Welsh industry, new markets were opened up at home and abroad, but the 1891 peak output was not exceeded until 1910.

For tin producers in general the demand for tin by the tinplate industry remained buoyant, since protection proved highly effective in allowing the US tinplate industry to become established. US tinplate production rose rapidly in the nineties from only 1,100 tonnes in 1891 to 115,000 tonnes in 1895 and 366,000 tonnes in 1900. This more than made up for the fall in UK production from 595,000 tonnes to 508,000 tonnes in 1900. The UK remained the largest producer until 1912, when output reached 862,000 tonnes, which would imply a tin consumption of between 12,000 and 13,000 tonnes, more than double consumption in 1979. Very little tinplate was produced outside the UK and US before the First World War, these two countries, with 90 per cent of world production in 1913, probably using about 27,000 tonnes of tin in tinplate. World use of tin in tinplate was about 24 per cent of total tin consumption, considerably less than the tinplate share over the last 60 years, excluding the war years.

Fluctuations and Trends in Tin Consumption

During the interwar years the UK remained the second largest consumer, the highest-ever consumption reaching 28,200 tonnes in 1928, with tinplate accounting for 12,100 tonnes. The UK was still the second largest tinplate producer, but with little sign of growth, peak production in 1937 being only 12 per cent higher than in 1912 (see Table 5.1).

Some increase in UK consumption resulted from the growth of the automobile and electrical industries. Details in the Tin Council's statistical series are insufficient to give a better breakdown of consumption. Total consumption fell by one-third in the depression years 1929–32, a similar, but more sudden fall occurring in the 1938 depression, much of it due to the fall in tinplate production. Tin used in tinplate fell from 13,500 tonnes to 8,600 tonnes in this recession, compared with a fall of only 2,600 tonnes in the previous depression.

After the Second World War, UK tin consumption recovered quickly from the 1943 low of 17,900 tonnes to 27,800 tonnes in 1947, which has proved to be the postwar peak. Consumption fell heavily in the 1948–9 recession to 21,100 tonnes, partly because of reduced industrial production, partly because of substitution in certain uses, notably in collapsible tubes and foil. But tinplate continued to use more tin until 1952, after which the switch to electrolytic tinplate more than offset the increase in tinplate production. By 1969 UK production had reached

Table 5.1: UK Consumption of Primary Tin Metal in Tinplate and Other Uses, 1925-1979 (000 tonnes)

	Tinplate		Other uses		Total
		%		%	
1925	11.3	45.5	13.5	55.5	24.8
1930	12.1	52.7	10.9	47.3	23.0
1935	10.2	46.7	11.6	53.3	21.8
1947	9.2	33.2	18.5	67.8	27.7
1950	10.0	42.2	13.6	57.8	23.6
1955	10.0	43.1	13.2	56.9	23.2
1960	11.5	50.0	11.7	50.0	23.2
1965	9.3	47.7	10.2	52.3	19.5
1970	8.0	47.0	9.0	53.0	17.0
1975	5.7	46.7	6.5	53.3	12.2
1979	5.9	52.9	5.2	47.1	11.1

Source: ITC, *Report on the World Tin Position*, 1965; *Tin Statistics*, 1965-75, 1969-79.

its postwar peak, with a tin consumption about one-third less than in 1952.

Since the early sixties, UK tin consumption has been falling more or less continuously, after a ten-year period of relative stability between 1954 and 1963. During that period electrolytic tinplate continued to replace hot-dipped tinplate, while in other uses of tin increases in consumption tended to offset reductions. In most of the sixties, and especially in the seventies, UK tin has been affected by the lack of growth in both tinplate production and UK industry generally. There has been no expansion of tinplate exports, in sharp contrast to the experience of most other industrial tinplate producing countries. In the last five years exports have been about one-third below the steady 360,000 tonne level between 1968 and 1971. Other European countries and Japan have surpassed UK exports, and only in *per capita* tinplate consumption has the UK remained ahead of these countries.

The fall in UK tin consumption by 1978-9 to its lowest level this century is only partly due to substitution. It is a striking indication of UK industrial experience in the seventies, that over the range of categories of tin use listed in the Tin Council's statistical series, there is none in which the UK used more in 1978 than in 1968. Although the data do not permit exact comparisons, it appears that no other major industrial country, even including the US, failed to use more tin in at least one category. It could be argued that the UK's industrial structure is simply changing in a way which involves less use of tin, but alternatively, for those who think that the UK is passing through a process of deindustrialization, the trend of tin consumption during the seventies might be regarded as further confirmation of this theory.

The USA

The American Attitude to Tin

The US has been by far the largest tin consumer throughout the whole of this century and totally dependent on imports the whole time, except to the extent that a national stockpile of formerly imported tin could be drawn upon for current consumption. For at least the last 40 years, major US users have been particularly active in seeking economies in the use of tin, encouraged at times by several administrations. Both business and government thinking have been influenced by dependence on imports. US policy in the thirties reflected opposition to the tin cartels. During the forties it was the risk of military conflict in southeast Asia which conditioned US policy towards tin, and later there were fears that the main suppliers were vulnerable to communist influences. Throughout the history of the postwar International Tin Agreement there has been an important strand of American thinking in business circles that the Agreement, in spite of its objectives, has worked at times rather like a cartel, and that, as the largest consumer, the US has been the main victim.[2]

The Effects of Tin-economizing Technology

Until Pearl Harbor, there was little evidence of actual substitution for tin in the US, although a number of firms were carrying on research into tin-economizing technology. Before the war, the US reached its maximum rate of consumption in 1929 at 86,400 tonnes of primary tin. This was about 50 per cent of world consumption. In the worst of the depression years the US continued to account for about half of world consumption, and absorbed an even higher proportion of world production if the strategic stockpile purchases are included. Consumption rose sharply during the first year of the Korean War to 72,200 tonnes, then fell to only 46,000 tonnes in 1952. After recovering to 61,000 tonnes in 1956, there was another sharp fall in the 1958 recession. Thereafter, consumption remained relatively stable until the 1974–5 recession, fluctuating around an average of 55,900 tonnes. In this recession, consumption fell to the lowest level since 1934. A subsequent recovery in the late seventies has still left consumption below the average of the sixties, and only about one-quarter of

Table 5.2: USA: Tin Consumption by Broad End-use Categories, 1968–1977 (000 tonnes)

	1968	1969	1970	1971	1972	1973	1974	1975	1976	1977
Cans and containers	27.4	25.4	23.8	22.5	20.4	21.1	21.5	17.6	9.8	17.8
Transportation	8.6	8.5	8.0	7.7	8.3	9.5	6.7	6.0	6.5	6.2
Machinery	8.0	8.4	7.4	6.4	6.7	7.6	6.4	5.7	6.1	5.7
Electrical	12.4	12.7	11.3	10.6	10.4	11.7	8.4	7.3	8.4	8.4
Construction	11.2	11.6	10.5	9.1	9.1	9.7	7.9	6.8	8.1	8.0
Chemicals	3.2	3.2	3.2	3.7	4.1	4.9	5.0	4.0	5.6	5.6
Other	3.2	3.0	2.6	2.0	2.2	2.4	2.4	1.7	1.9	1.7
Total consumption	74.0	72.8	66.8	62.0	61.2	66.9	58.3	49.1	56.6	53.6
Total primary consumption (industrial demand less secondary)	60.4	58.9	54.7	50.7	49.8	55.0	46.6	39.9	46.5	42.1

Source: *Tin International*, January 1980. Statistics from US Bureau of Mines, *Minerals in the US Economy: Ten-Year Supply-Demand Profiles for Non-Fuel Mineral Commodities (1968–1977)*, Pittsburg, 1979.

world consumption. Both the US and UK, therefore, now consume much less tin than they did before the last war.

According to the pattern of industrial uses by broad categories, there has been a general fall in US consumption since 1968, except in chemicals, as Table 5.2 shows. The modest increase in chemicals has offset only the decline in machinery. Tin consumption for cans and containers, although coating a larger number of units per tonne if measured by area of tinplate used, has fallen heavily. To a large extent this is due to the highly competitive nature of the American container market for food, beverages and other packaged goods. Although two large companies, American Can and Continental Can, dominate the US metal container industry, they compete vigorously with each other, as well as with other firms producing metal containers. Both large companies spend heavily on experiments with different materials for can-making. The large steel companies are equally active in the search for steel-based can-making material, which will allow them to maintain a large output in a profitable part of the steel industry.

US consumption of tin for tinplate has fallen well below the levels of the fifties (see Table 5.3). Consumption for other purposes has benefited from the expansion of certain industries, which have tended to check a secular decline. The result is that the relative importance of the tinplate use of tin in the US, after rising appreciably in the interwar years, and still more in the forties and fifties, has again fallen below the 40 per cent level. For a number of years in the forties and fifties the share of tinplate was over 60 per cent. As late as 1962 it was 62 per cent. Since then there has been a progressive fall to less than 40 per cent in the period 1977 to 1979. In 1978 it was lower than in any year since 1930. Since the US is still the largest user of tin, it is worth considering the possible effect of the decline in the share of tinplate. Most tinplate is used for food and beverage cans, consumption of which is a function of consumer incomes. Knorr found from the experience of the US in the interwar years that the consumption of canned food fluctuated in close relation to consumer income and to a somewhat greater extent than total food consumption.[3] Whether fluctuations in other uses of tin are greater than in tinplate depends on the importance of capital goods as a source of tin demand. American experience suggests that non-tinplate uses tend to be more unstable than tinplate usage. If this is a pointer to the future, a relative decline in the tinplate share of tin consumption would make the tin market more volatile than it has been for several decades. On the other hand, there might be some advantage for tin in depending less heavily on a single

Table 5.3: USA: Consumption of Primary Tin, 1925–1979 (000 tonnes)

	Tinplate	%	Other uses	%	Total
1925	25.2	32.7	50.0	67.3	75.2
1930	28.2	36.2	50.1	63.8	78.3
1935	27.7	43.7	35.8	56.3	63.3
1947–8	31.6	52.5	28.7	47.5	60.3
1952–3	29.8	49.1	30.6	51.9	60.4
1957–8	31.1	59.8	20.7	40.2	51.8
1962–3	28.9	52.3	25.1	47.7	54.0
1967–8	29.2	50.0	29.2	50.0	58.4
1972–3	21.5	38.0	35.0	62.0	56.5
1977–8	17.9	37.3	30.1	62.7	48.0
1979	17.9	36.5	31.1	63.5	48.0

Source: ITC, *The World Tin Position*, 1965; ITC, *Tin Statistics*, 1965–75, 1969–79.

major source of demand where major consumers spend heavily on ways of economizing tin, and where it is subject to intense competition from substitutes.

Continental Western Europe

Pre-war Tin Consumption

Before the Second World War tin consumption in the six countries of the original European Community was disproportionately smaller than UK consumption (see Table 5.4). In spite of the greater size of its economy and a population of 65 millions compared with the UK's 46 millions, prewar Germany used only about half the UK level. This disproportionately small consumption in Germany and the other advanced countries of Western Europe was to a large extent due to a smaller volume of tinplate production. The combined output of Germany, France and Italy, the only other European producers of tinplate in 1937, was only half the British output. All the Continental countries had a much lower preference for canned food than the UK. It is also likely that Germany was used to greater economy in materials consumption than the UK, a practice which would certainly have been encouraged by the Nazi regime as part of its self-sufficiency policy. As a comparatively under-industrialized Western economy before the war, France might have been expected to use less tin than the UK. French tin consumption has been remarkably steady during

Table 5.4: European Community: Consumption of Primary Tin Metal, 1925-79 (000 tonnes)

	German Federal Republic[a]	France	Italy	Belgium	Netherlands	Total
1925	10.0	11.0	4.2	1.3	1.0	28.6
1930	13.7	11.6	4.3	1.6	1.2	32.4
1935	10.6	8.3	6.6	1.3	1.2	28.0
1950	7.9	7.8	2.7	1.4	3.1	23.2
1955	8.3	9.7	3.0	2.1	2.6	25.7
1960	(11.0)	11.2	5.0	2.8	3.1	33.1
1965	11.8	10.3	5.8	2.4	3.4	33.7
1970	14.1	10.5	7.2	3.0	5.5	40.3
1975	12.0	10.3	6.3	4.4	3.6	36.6
1979	13.7	9.7	6.0	2.4	4.8	36.6

Note: a. Prewar Germany. The 1960 figure for Germany is an estimate since figures for the years 1959-61 were distorted by special circumstances arising out of a temporary trade in secondary tin alloys.

Source: ITC, *Report on the World Tin Position, 1965*; *Tin Statistics*, 1965-75, 1969-79; and *Statistics of Tin*, 1945-70.

normal peacetime conditions over the last 55 years, which suggests that the growth of tin-using industries has more or less kept pace with tin-economizing technology.

Post-war Changes

After the war, consumption in the Community group grew only slowly, not reaching the prewar peak until 1959–60. Tinplate production tripled by 1960, but the electrolytic process kept down the amount of tin required. Continued changes in consumer habits led to a 150 per cent increase in tinplate production between 1960 and 1979, with no more than a 12 per cent increase in tin consumption. Total tin consumption grew by less than one-quarter between 1960 and the late seventies, which was not enough to offset the sharp fall in the UK consumption. The only other significant West European consumer is Spain, where consumption rose from less than 1,000 tonnes in the fifties to 4,500 tonnes in the mid-seventies, since when further expansion seems to have been checked by a lower average tin coating on an expanding tinplate output.

Japan

Post-war Economic Growth and Tin Consumption

The great expansion of the Japanese economy since 1945 has made it the second largest tin consumer. Peak Japanese consumption, so far, was 38,700 tonnes in 1973, which was only about 12 per cent less than US consumption in 1975, admittedly not a year of strong demand in the US, but indicative of the narrowing gap between the volume of metal-using industries in the two countries. In 1979 Japanese consumption was about two-thirds the American level, and nearly equal to the combined consumption of the UK, West Germany and France. Thus Japan has become, in tin as in other industrial materials, a major factor in the world commodity markets.

As in most other countries, Japan's chief use of tin is for tinplate, of which it is the second largest producer and easily the largest exporter (see Table 5.5). On the other hand, Japanese consumption of tinplate is still less than that of the UK. Most Japanese tinplate is now electrolytic, but the average tin coating is higher than in the major Western countries. This is certainly not due to any backwardness in adapting the new technology, but rather to differences in the type of canned products, some of which require a thicker tin coating than

Table 5.5: Japan: Consumption of Primary Tin Metal in Tinplate and Other Uses, 1925-1979 (000 tonnes)

	Tinplate		Other uses		Total
		%		%	
1925	–	–	–	–	3.3
1930	–	–	–	–	4.1
1935	–	–	–	–	6.2
1950	1.2	25.5	3.5	74.5	4.7
1955	3.3	50.0	3.3	50.0	6.6
1960	5.1	38.5	8.1	61.5	13.2
1965	8.4	48.3	9.0	51.7	17.4
1970	10.6	43.0	14.1	57.0	24.7
1975	11.6	41.3	16.5	58.7	28.1
1979	12.4	40.0	18.8	60.0	31.2

Source: ITC, *Report on the World Tin Position*, 1965;*Tin Statistics*, 1965-75, 1969-79.

others. Japan has been in the forefront of research in developing alternative coatings for tinplate, and made the first economic breakthrough with a chrome coating.

Importance as a Tin Consumer

The important position now occupied by Japanese producers in the electrical and electronic industries naturally suggests a high consumption of tin in alloys, which is borne out by solder consumption. In 1979 some 10,200 tonnes of primary tin were used for solder, compared with 2,300 tonnes in West Germany, about 1,000 tonnes in the UK, and 13,000 tonnes in the US, plus 5,000 tonnes of secondary tin.

The growth of Japanese consumption has compensated for the long-run decline in US and UK consumption. Out of an increase in world consumption of 24,000 tonnes between 1960 and 1979, Japan accounted for 16,000 tonnes, some 2,000 tonnes more than the fall in US and UK consumption. During the seventies, however, there were signs of a marked slowing down in the growth of Japanese consumption. Average consumption in 1971-2 was only slightly higher than in 1978-9, whereas in the five years up to 1970 there had been a 50 per cent increase. Tinplate production grew much more slowly in the second half of the seventies, largely due to a less buoyant export market.

It seems likely that the rise in the relative price of tin has had some effect on Japanese demand. Dependence on imported raw materials in general is now so great that there must be a strong incentive to make use of any economically practicable ways of economizing on materials consumption. The incentive might be expected to be greater than in the early seventies, since Japan has been particularly vulnerable to the rise in the price of energy. It would be logical to try to reduce the import content of high-priced metals in manufacturing output, with tin being an obvious target. It was hardly surprising that Japan was the first country to substitute chromium-coated steel for tinplate.

Fluctuations in Demand and Prices

Like other industrial materials, most of which are consumed in the industrial market economies, tin has always been vulnerable to fluctuations in business activity, and has a long record of market instability going well back into the nineteenth century. Historical statistics from the period when demand for tin began to expand rapidly during the third quarter of last century show a high degree of instability in both volume and price.[4]

In the interwar years the tin market was highly susceptible to fluctuations in US demand for current consumption and stocks, since the US accounted for about 50 per cent of world consumption. In the great depression of the early thirties, US demand fell from a peak of 86,300 tonnes in 1929 to a low of only 36,100 tonnes in 1932, followed by a sharp recovery to 60,700 tonnes in 1933, chiefly because of the expansion of demand for tinplate by the food industry. The fall in US demand amounted to two-thirds of the fall in world demand, compared with the 50 per cent share of normal demand. In the 1937–8 fluctuation, US demand again accounted for a disproportionate share of the contraction in world demand. Out of a 30,500 tonne fall in demand, the US accounted for 24,400 tonnes. In the subsequent recovery, the US accounted for 18,300 tonnes out of a 20,300 tonne increase in world demand. In this cycle there was a violent fluctuation in both the US and UK tinplate demand for tin, but a small, steady increase in demand by the rest of the world, which might have been influenced to some extent by war preparations in National Socialist Germany.

During the earlier postwar years the US was again the main source of fluctuations in demand for tin. In the 1948–9 recession world demand fell by 15,200 tonnes, US demand by 12,900 tonnes. In the 1957-8 recession, however, demand fell only in the US and UK. By the early seventies the US had a considerably smaller share of world demand, whereas the shares of Japan and the European Community had risen to 18 per cent and 30 per cent, respectively. The US share was down to 30 per cent, compared with 40-50 per cent in the fifties. In the 1974-5 recession, the most severe since 1938, the absolute fall in Japanese demand was almost as great as that of the US, and greater than that of the entire European Community. Compared with 1973, Japanese demand fell by 14,000 tonnes, US demand by 15,000 tonnes, and European Community demand by only 9,000 tonnes.

It seems that the relative increase in the Japanese share of world consumption has not contributed to greater stability in the world tin market. If the experience of the 1974-5 recession is a guide, the growth of Japanese demand for both tin and other industrial materials has simply created a new source of instability. A sharp reduction in Japanese demand for imports, with some destocking, had a significantly adverse effect on the markets for tungsten, nickel, copper, aluminium and molybdenum, as well as tin.

Tin Council statistics of apparent world consumption show a sharp increase in 1973 to the highest peacetime level, from 192,000 tonnes to 214,000 tonnes, followed by a fall to only 174,000 tonnes in 1975. The upsurge in demand, which probably reflected consumer stock changes, as well as current consumption, was met by buffer stock and US stockpile sales. The fall in world demand during the recession was comparable to the fall in the 1938 depression, but much less, of course, than in the great depression of the early thirties.

Fluctuations in demand are usually associated with large fluctuations in market prices. The record of prices in the last 30 years of the nineteenth century shows that the tin market suffered from acute instability. Between 1868 and 1872 the average annual price rose by 55 per cent and fell by 50 per cent in the next few years. The average to-year fluctuation in the 1880s was 15 per cent, ranging from a low of 2 per cent in 1891-2 to 58 per cent in 1898-9. The interwar years set new records for price instability, with the postwar boom and slump, the slide from the peak price of £321 a tonne in 1926 to £100 a tonne in 1931, and a brief, but violent fluctuation from £311 in 1937, the best year of the thirties, to £153 in 1938. The average year-to-year change between 1921 and 1939 was 18 per cent. Since the Second World War there have been several periods of violent instability, especially during the Korean War and the major commodity boom and recession of the first half of the seventies. The LME price rose by over two-thirds between 1949 and 1951, then fell by one-third by 1953.

Between 1972 and 1974 the LME price rose by 134 per cent and fell by nearly 20 per cent in 1975. With currency divergences between sterling and the Malaysian dollar, however, the 1972–4 increase in Penang was only 80 per cent and the 1975 fall 15 per cent.[5]

Fluctuations in current consumption have also been reinforced at times by inventory fluctuations. Unfortunately, accurate information on stocks held by industry and merchants is unobtainable, a deficiency recognized as the weakest link in the Tin Council's statistical picture of the world market. Knorr gave some consideration to inventories in the US during the interwar period, quoting official estimates that tin metal stocks carried by industrial consumers between 1928 and 1940 varied from a yearly average of 2,800 tonnes to one of 6,300 tonnes, compared with a fluctuation in estimated consumption from 40,600 tonnes to 87,000 tonnes. According to Knorr, the ratio of stocks to consumption was at its highest of 11 per cent in 1931, but during the 12-year period it was 6.6 per cent on the average, or less than one month's requirements.[6]

Postwar consumer-held stocks in the US, reported to the Tin Council, seem to be higher than those quoted by Knorr for the interwar years when consumption was higher, but it is uncertain whether the two series are comparable. What is striking about US consumer-held stocks is the very marked decline from the mid-sixties to the late seventies, as shown in Table 5.6. Whereas they were equal to 23 weeks' consumption in 1965, they were down to only 4 weeks' consumption in 1979. All metal stocks reported by leading countries, excluding stocks in transit, were also somewhat lower in the late seventies than in the fifties and sixties, perhaps due to higher tin prices and interest rates. Another reason for the fall in US consumer stocks might have been the belief that stockpile releases could be relied upon to compensate for a shortfall in current production.

The World Bank study incorporated 'implied stocks' in its model, while emphasizing that the lack of accurate information from industry and trade made it impossible to analyse in detail the behaviour of inventories. Its conclusion was that 'among the many market forces which affect the price of tin, two factors, namely the change in inventories and inflation, explain most of the variation in price'.[7] It also found that the estimated price equations in the tin model confirmed the expectation that there should be an inverse relationship between prices and the ratio of inventories to demand or supply.

Fox has pointed out the belief that inventory variations in the interwar years influenced the industry's thinking on the possibility of

Table 5.6: World Tin Stocks,[a] 1965-1979 (000 tonnes)

	Tin-in-concentrates[b]	Tin metal	(with US consumers[c])	Total	As per cent of consumption
1965	10.0	44.7	26.1	54.7	33.3
1966	15.6	41.6	20.9	57.2	34.7
1967	12.4	45.5	17.6	57.9	35.0
1968	13.8	55.2	16.1	69.0	41.0
1969	11.2	42.1	12.7	53.3	30.0
1970	12.6	36.2	9.6	48.8	28.0
1971	9.6	39.2	8.0	48.8	27.7
1972	8.4	41.7	8.5	50.1	28.0
1973	12.1	38.0	7.6	50.1	25.0
1974	10.3	39.9	9.2	50.2	26.9
1975	15.7	43.5	7.5	61.2	38.0
1976	7.8	38.7	6.2	46.5	25.9
1977	8.4	35.9	6.8	44.3	26.9
1978	9.9	33.8	4.8	43.7	24.1
1979	10.2	26.6	4.0	36.8	21.5

Notes: a. Some secondary tin metal is included. Excludes buffer stock tin. b. Stocks held at mines, in transit and at smelters. c. Excludes stocks held by jobbers and importers. These are included in the total.

Source: ITC, *Tin Statistics*, 1965–75, 1969–79.

a successful buffer stock, which was eventually incorporated into the prewar and postwar agreements.[8] It will be clear from a later chapter, however, that the producers were reluctant to accept a large enough buffer stock, able to deal effectively and without recourse to any restrictions on exports or output, with serious fluctuations in both commercial stocks and current consumption. Accurate and up-to-date information about the behaviour of stocks would greatly help the Tin Council in formulating policy for both the buffer stock and export control. In postwar years the problems caused by variations in stocks have been complicated by vacillations in US stockpile policy, in addition to those resulting from other market influences.

Notes

1. The impact of the McKinley tariff is discussed by Minchinton, *The British Tinplate Industry*, Clarendon Press, Oxford, 1957. See also a series of articles on the history of tinplate by Anthony Smith in various issues of *Tin International* in 1977 and 1978.

2. After several years' membership of the ITA, American business opinion remained unchanged. In April 1980, before the opening of negotiations for a sixth agreement, the attitude of leading US users was voiced by a representative

of the American Iron and Steel Institute: 'We cannot support the participation of the US in the sixth tin agreement. We do not believe that the US has realised any economic benefits though its membership.' Quoted in *Tin International*, March 1981.

3. Knorr, *Tin under Control*, p. 45.

4. For historical prices see Schmitz, *World Non-Ferrous Metal Production and Prices 1700–1976*, Cass, London, 1979.

5. Currency changes have led to different rates of change in tin prices in national currencies over the past ten or twelve years. The decline of the foreign exchange value of the pound sterling up to the late seventies caused the sterling price of tin to rise much more than the German mark or Malay dollar price. In 1973, for example, the mark price was about 20 per cent less than in 1965, and the Malay dollar price about 3 per cent less, whereas the sterling price was about 25 per cent and the US dollar price about 20 per cent higher. A large difference also appeared in the late seventies between the US dollar price and the prices in other currencies. After the currency upheavals of 1971, the ITC price range was expressed in Malay dollars from July 1972.

6. Knorr, op. cit., p. 30.

7. World Bank, *The World Tin Economy: An Econometric Analysis*, p. 24. In the report, implied stocks are defined as 'stocks in the previous period plus total market availabilities minus total market disappearance in the current period'.

8. Fox, *Tin: the Working of a Commodity Agreement*, p. 107.

6 SECONDARY TIN

Like other metals, a certain proportion of tin consumed consists of tin which has previously been recorded as primary or virgin tin. This reused or secondary tin does not appear to any great extent in the form of metal, since most secondary tin remains in the alloys in which it has been combined in various proportions with other metals. Only a small percentage of secondary tin competes directly with primary tin. Accurate information about the supply and consumption of secondary tin in one form or another is not available. The main source of information about the consumption of secondary tin is the US, which has recorded statistics of secondary tin for many years (see Table 6.1), but even here the figures must be treated with some reserve.

Secondary Tin Metal

Secondary tin metal comes from tin recovered from the tinplate mills and the can-making plants which have large amounts of clean scrap in the form of unused tinplate and cans. The recovery of tin metal from clean scrap is a long-established economic operation, but the total amount of metal recovered in this way is now much less per tonne of scrap treated than it was when the only tinplate was hot-dipped. The recovery rate in the US in 1947, when there was still over 50 per cent hot-dipped tinplate among the scrap, was 20.05 pounds of tin per tonne of scrap; by 1962 it was down to an average of 10.06 pounds when most scrap was electrolytic. The total amount of tin metal recovered, however, did not change much, since the amount of scrap increased. Assuming an average recovery rate of 97 per cent of the tin coating by the detinning plants, the 1979 figure was probably just under 9 pounds. A continued reduction in the tin coating was partly compensated by an increase in the area of tinplate covered by a given quantity of tin as a result of the use of a thinner steel base. Since the late sixties, a fall in tinplate production and in the amount of solder used for can-making has reduced the amount of secondary tin metal recovered in the US. Consumption of secondary tin metal in the US has been falling since the late sixties, but with the fall in primary tin metal consumption over the same period, the proportion of secondary has remained within the usual 4-5 per cent range.

Table 6.1: USA: Consumption of Primary and Secondary Tin, 1965-1979 (000 tonnes)

	Primary tin		Secondary tin[a]		Total
		%		%	
1965	59.5	69.7	25.9	30.3	85.4
1966	60.3	70.1	25.7	29.9	86.0
1967	58.9	71.7	23.2	28.3	82.1
1968	59.9	71.8	23.5	28.2	83.4
1969	58.3	71.6	23.4	28.4	81.7
1970	53.9	72.0	21.1	28.0	75.0
1971	52.8	74.3	18.3	25.7	71.1
1972	54.4	77.5	15.8	22.5	70.2
1973	59.1	78.6	16.8	21.4	75.9
1974	52.4	79.8	13.3	20.2	65.7
1975	43.6	78.2	12.2	21.8	55.8
1976	51.8	82.2	11.2	17.8	63.0
1977	47.6	78.5	13.1	21.5	60.7
1978	48.4	79.7	13.1	20.3	61.5
1979	49.5	79.3	13.0	20.7	62.5

Note: a. In metai and other forms.

Source: ITC, Tin Statistics 1965-1975, 1969-1979. Data drawn from Tin, Mineral Industry Surveys, US Department of the Interior, Bureau of Mines.

A Tin Council estimate of world consumption of secondary tin metal in 1978-9 was about 14 per cent less than in 1968-9. Only about 4 per cent of world metal consumption in 1978-9 consisted of secondary tin metal. Of the tin used in tinplate about 10 per cent is recycled, the other 90 per cent is thrown away by consumers in the form of discarded tin cans; thus about one-third of world tin production is used only once and disappears into rubbish tips.

From the point of view of the world as a whole this is a serious sacrifice of a relatively expensive non-ferrous metal. It must be admitted, of course, that the development of ever-thinner tin coatings has greatly reduced the waste of tin in this way, while preserving the very great benefit conferred on modern civilization by a safe method of long-life food preservation. But in the last decade or so, to the normal commercial incentive of economizing in the use of an expensive metal has been added the pressure to clean up the environment. The tin can is an obvious target of this campaign. Hence, for environmental reasons, there has been an intensification of research into commercially successful ways of recovering both tin and steel from old scrap tinplate.

The emphasis lies on 'commercially successful'. In several industrial countries projects have been under way for years to recycle the tin in used cans, but although considerable progress has been made in both the US and UK, it has proved exceptionally difficult to extract the tin profitably from the cans. The US Bureau of Mines Secondary Resource Recovery Group reported in 1978 that 'nowhere in the US today is municipal refuse being detinned on a commercial basis'. It concluded regretfully, 'despite the extensive network of can reclamation plants in the US and the most sophisticated detinning technology, the economic viability of reclaiming tin from waste cans has still to be established'.[1] A Dutch authority in 1980 confirmed this judgement in referring to a new project for detinning used cans: 'The various parties involved in this project fully realised, from the start, that an operation in which exclusively tinplate is being separated, could never develop into a paying proposition'.[2] The most promising British project was launched in February 1980. It was believed that on a minimum economic throughput of 100,000 tonnes of waste cans it should be possible to produce about 40 tonnes of high-purity metallic tin, plus a large amount of steel. By November 1980 it was reported that 'at today's prices the Materials Recovery Limited plant is not economically viable'.[3]

Success, therefore, seems to be elusive. Meanwhile, the producers of tin cans are under attack from environmentalists and resource conservationists as a sizeable part of world tin production is irretrievably lost each year. One speaker at the Second International Tinplate Conference in 1980 argued in favour of industry support for can reclamation on grounds of the social benefits, accepting that at least under present conditions it could not be made to work on purely commercial terms.[4] It might be argued also that a better case could be made for a public subsidy to reclamation plants than for some other objects of public subsidy, in view of the scarcity value of the sacrificed tin metal. There might, after all, be a favourable effect on the price of primary tin to the consumer if the supply of secondary tin metal could be increased. It would be the equivalent of opening up new mine capacity.

Secondary Tin in Alloys

With the primary tin which goes into alloys such as solder, brass and bronze, the position is very different. These alloys can be recycled again and again as the scrap which contains them becomes available to new users, but their tin content does not reappear as metal. Although in some alloys the tin content may be very small, a large amount of secondary tin is contained in the volume of alloys which has been produced over the years. Most of the stock will be currently held in the industrial countries which have been the main consumers of primary tin. All the industrial countries, including those with no primary tin resources, share this stock of secondary tin in alloys, which has been accumulated in a great variety of manufactures. As these products become obsolete, or for some reason are scrapped, some proportion of the tin-containing scrap will reach the recovery plants and the alloys recovered will be available for further use. Most secondary tin in alloys comes in bronze, brass and solder, according to US data. This is probably also true of other industrial countries, as appeared from a Tin Council inquiry in the early sixties into secondary tin consumption.[5]

The rate at which scrap is processed so that the alloys can be recycled is uncertain, since not enough is known about the sources of supply in many countries. It has been estimated that the typical life span of important brass mill products in the US varies from 4 to 29 years, with recovery rates varying from 12 per cent to 44 per cent.[6] An uncertain amount of tin would be contained in the scrap. So far as copper is concerned, US estimates suggest that 'of the total volume of obsolete copper products becoming available every year, 30 per cent is recovered; 1 per cent is dissipated; 21–35 per cent finds its way into solid waste disposal sites, and 34–48 per cent is unaccounted for'.[7]

Figures for the US suggest that the amount of secondary tin in alloys has fallen since the forties and fifties, when consumption ranged between 24-32,000 tonnes. This was equivalent to about 35 per cent of total US tin consumption, and about 50 per cent of non-tinplate uses of tin. In some years the tin content of reused alloys considerably exceeded the amount of primary tin in new alloy production. During the seventies, US consumption of secondary tin in alloys fell sharply, apparently as the result of a fall in bronze and brass consumption, in which secondary tin is usually more important than primary tin. In 1978–9 the relative proportions in bronze and brass were 6,000 tonnes of secondary to 3,000 tonnes of primary tin. In 1958–9 the corresponding figures were 12,000 long tons and 3,500 long tons.

The question naturally arises of the relationship between the price of primary tin and the supply or consumption of secondary tin, an issue which was examined by Knorr for the US in the interwar years. His findings were that a comparison of the course of tin prices and secondary tin production in the US from 1922 to 1940 reveals that 'the recovery of secondary tin is markedly responsive to changes in the price of virgin tin, but this responsiveness operates only within a certain range'.[8] Knorr maintained that the range is not constant, but that competition between secondary tin is limited by 'definite conditions operating at any one time'. In his opinion, these conditions were the capacity of the reclamation plants, the availability of scrap and the cost of recovery, but the range of the competition was never wide enough to check a major rise in the price of tin in the interwar years.

A major attempt was made by the Tin Council to obtain a range of information on the production and consumption of secondary tin, as well as an answer to the question – how far is the movement in the price of primary tin metal a factor which directly affects the consumption of secondary tin?[9] Another question concerned the hypothesis that a mass of secondary tin in a country might constitute a kind of pool of tin upon which it could fall back in an emergency. There seemed to be general agreement that the second question was not really relevant, since this tin was 'locked up' in alloys and, being inseparable from them, only available when the goods containing them were scrapped. Whether these goods should be scrapped in an emergency to use the alloys for some other purposes would depend on their relative importance.

Answers to the key question about prices and consumption were mixed. The UK reply stated:

> although the prices of primary and secondary tin are closely linked and move in conjunction with one another, movement in the price of primary tin does not seem to have much influence on the consumption of secondary tin in the UK unless the price of primary tin happens to be exceptionally low or extremely high . . . The price of primary tin would have to be at a very low level to cause any significant interruption in the flow of secondary tin coming forward for treatment . . . production of secondary tin is continuously at a high level in relation to the secondary materials available for treatment, and therefore there would seem to be little scope for increasing the output of secondary tin.[10]

The Japanese view was that the movement in the price of primary tin had hardly any influence on the consumption of secondary tin. According to the US reply: 'the volume of all secondary metals fluctuates with the price of primary metals', but 'the value of lead, tin, antimony, copper, and other elements in the scrap should all be taken into consideration'.[11] A rise in the price of tin, therefore, when the tin content is small, might have no effect on the recycling of the alloys. Other replies which conceded some link between the price of primary tin and the supply of secondary tin did not give it much importance.

It is interesting that the World Bank study did not include the production of secondary tin on the supply side of its model of the world tin economy. Commenting on the omission, the study referred to the paucity of reliable data or complete absence of published data in many countries. 'For countries for which such data could be obtained [e.g. the United States], they indicate little variation in total supply. Secondary tin production in the United States failed even to respond to the high tin prices of 1973 and 1974'.[12] The Bank study suggested that 'an explanation for this apparent lack of price responsiveness can probably be found in the increase in recovery costs that took place over the same period', and pointed out that 'attempts to capture the price response of the US supply of secondary tin statistically were largely unsatisfactory'.[13]

Notes

1. *Tin International*, January 1978.
2. *Tin International*, November 1980, Report on the Second International Tinplate Conference, London, October 1980.
3. Ibid.
4. Ibid.
5. ITC, *Report on the World Tin Position*, 1965, Ch. 5.
6. R.F. Mikesell, *The World Copper Industry: Structure and Economic Analysis*, Baltimore, John Hopkins Press and Resources for the Future, 1979, p. 343.
7. Mikesell, op. cit., p. 345.
8. Knorr, *Tin under Control*, p. 40.
9. ITC, op. cit., pp 154-6.
10. UK answer to the ITC questionnaire on secondary tin. This and other answers were the basis for Ch. 5 of the 1965 ITC report.
11. US answer to the ITC questionnaire.
12. World Bank Staff Commodity Paper No. 1, p. 11.
13. Ibid., p. 11.

7 SMELTING AND THE MARKETING SYSTEM

Since the majority of tin-producing countries consume very little of their output, most tin enters into international trade and reaches the ultimate industrial consumers through a complex of dealers and markets. Some tin is marketed by supply contracts between governments or their state corporations and consumers. State-produced Indonesian tin is sold by the state corporation through offices in London and Antwerp. Thailand's output is channelled to consumers through one firm of merchants. The Malaysian Mining Corporation, the world's largest tin-producing organization, with a substantial state interest, sells direct to customers and through a London agent. Some Bolivian tin is sold direct to the Soviet Union. The American General Services Administration sells surplus stockpile tin by competitive bids. A substantial amount of tin is handled by tin traders or merchants between the Penang smelters and the consumers.[1]

For this rather complex pattern there are several reasons. Firstly, the tin mining industry is very fragmented, with a few large organizations, both private and public, a large number of smallish units and a very large number of extremely small producers. Secondly, there is a limited number of smelters, of which the largest do not have captive sources of tin concentrates for smelting. The important role of the large Malaysian smelters in the marketing system is described later in this chapter. The nationalized Bolivian and Indonesian producers sell most of their output to state-owned smelters. Thirdly, there is a very diversified demand for tin, with a relatively small number of large users in the steel and automobile industries, and a large number of small users. It is to be expected that the small users would buy from intermediaries, the dealers in tin who buy from private smelting companies or from state organizations. The large steel companies also obtain their tin indirectly, and do not seem to be interested in forming financial links with producers.

Smelting

The two Malaysian smelters belong to old-established companies dating back to the late nineteenth century, when British enterprise and capital replaced the small-scale smelting operations of the Chinese miners who had dominated the Malaysian industry for several decades. Both smelters were established in the 1880s. Thailand's first smelter was not built until 1965. The Dutch East Indies had smelting facilities before the Second World War. After the war, Indonesian concentrates were exported for smelting until new plant was commissioned by the state mining company in 1967. Bolivia did not get its first smelter until 1971, and part of its output is still exported for smelting. Local smelting is also carried on by the other leading developing country producers, Nigeria, Zaire and Brazil. As far back as 1930, two-thirds of non-communist production was smelted in developing countries.

Before the last war, West European smelters handled one-third of non-communist output. All Nigerian and most Bolivian output went to the UK; Belgium handled concentrates from the Congo and Rwanda; a Dutch smelter took part of East Indies output. The largest consuming country, the US, had no smelting facilities of its own. An attempt had been made as early as 1903 to set up a tin smelting plant in the US, but it was frustrated by the imposition of a 40 per cent *ad valorem* export duty on concentrates from British Malaya to destinations outside the British Empire.[2] Since Bolivian concentrates, which were to be treated by the proposed smelter, could not be treated economically without an admixture of purer concentrates from south-east Asia, the American smelter never began production. During the First World War, some tin smelting was started in the US when shipping problems affected the export of Bolivian concentrates to the UK. This continued on a diminished scale after the war, finally petering out in 1924 when Bolivian concentrates were diverted to Europe.

Smelting was not restarted in the US until 1942 when the Texas smelter was built to treat imported concentrates from several sources for both current consumption and the strategic stockpile. The Texas smelter now treats only a small proportion of US consumption, since smelting has increased in the producing countries. There has also been a contraction in the West European smelting industry for the same reason. The Dutch smelter at Arnhem has closed, as has one of the British smelters, suffering from technical problems, which compounded the effects of falling supplies of concentrates. West Europe now smelts less than 10 per cent of its tin consumption, and the industrial countries as a whole, including Australia, less than one-fifth of non-communist world production.

Although the cost of smelting has never been a high proportion of the selling price of tin metal in most producing countries – less than

2 per cent in Malaysia – developing countries became increasingly reluctant after the last war to lose any potential value-added from their mineral production. In the changing postwar climate it was logical that they should want as much control as possible over their basic industries. Bolivia in particular had cherished for many years the ambition of smelting its ores and reducing the transport cost from the mines through a foreign country to the nearest port and then thousands of miles by sea to the smelter. Bolivia, in fact, was paying for the shipment of a substantial amount of useless material to Europe, since a large part of the total tonnage is removed when the concentrate is smelted. Bolivia has achieved its objective of smelting at least part of its output, but costs are considerably higher than in other smelters because of the absence of accessible coal deposits, and particularly the difficulty of smelting its complex and low-grade ores in a smelter suitable for treating medium- to high-grade ores. Although precise data are not available, the so-called 'realization costs', which include smelting, transport, insurance, assays, weighing and penalties, are approximately 24 per cent of the total value of low-grade concentrates assaying 27 per cent tin and approximately 18 per cent of the total value of high-grade concentrates assaying above 50 per cent tin.[3] Smelting costs, however, are expected to be reduced with the completion of another smelter for treating Bolivian ores. The remaining British smelter still treats Bolivian and other ores, because it is capable of smelting economically low-grade, medium-grade, high-grade and complex ores. British smelting output, however, is now far below the level of prewar years. As much as 58,000 tonnes, over one-quarter of world production, was smelted in 1929; the 1978 output was down to 7,000 tonnes.

The closure of smelters in the industrial countries has not meant any contraction in world smelting capacity. Total non-communist capacity at the end of 1977 has been estimated at around 330,000 tonnes, compared with metal production of 203,000 tonnes in 1979. Another 80,000 tonnes of smelting capacity may be in the Soviet Union and China, and new smelting capacity may be installed in other countries. The result will be that 'the 1980s are likely to see a consolidation of the southeast Asian share of world tin metal output . . . a continuing contraction in Europe, and a continuation of the trend towards full domestic smelting in the tin producing countries'.[4]

The Marketing of Tin: Penang and the LME

Apart from smelting concentrates produced in Malaysia and elsewhere, the two Malaysian smelters play a major role in the marketing of tin metal. The smelting companies buy direct from the larger Malaysian mines and from a number of agencies who collect on their behalf the concentrates produced by scattered smaller mines. The small mines in more remote areas and the dulang washers sell their output to independent tin ore buyers, who re-sell it to the smelting companies' agents. The smelting companies then act as intermediates for the miners and the buyers of metal on the Penang market, one of the three markets on which the price of tin is determined, the others being the London Metal Exchange and the New York market.

Although far apart physically, the three markets are interdependent, with a close correspondence between them in the pricing of tin, allowing for freight charges, insurance and the effects of exchange rate fluctuations. There can be occasional divergences, since the LME price is more volatile than the Penang price. Unlike the LME, the Penang market is a purely physical market for tin, where the actual commodity is bought by traders and consumers. A large proportion of LME transactions consists solely of paper transactions which involve no actual movement of tin. Fluctuations in the LME prices are partly the result of speculative activity, which can be intense at times, especially during periods of exchange rate instability such as occurred in the seventies.

There is a marked difference in the methods of operation of the Penang market and the LME. On the former, the procedure is as follows. Buyers make confidential written bids for a certain amount of tin on a particular day. Their decision on the offer price is influenced by their view of market trends and market sentiment throughout the world, and by their need for tin within a particular time. The mines offer so much concentrates daily to the smelting companies, which act as the intermediaries after smelting the concentrates. The buyers' bids are listed at the buying prices on offer. (The smelting companies act in concert.) If the sum total of the bids exceeds the amount of metal offered for sale, a cut-off price is established at the point where the bids are equal to the supply. All bids above the cut-off price receive contracts in full. Those bidding at the marginal price where supply just equals demand are rationed to the extent of the balance after bids above the prices are met in full. Successful bidders are not allowed to resell their tin on the Penang market, nor are hedging facilities available, but buyers may hedge on the LME. The price determination process is

Table 7.1: Summary of Bids Received on the Penang Tin Market

Tonnes	Price M$ per pikul	Cumulative tonnage bid
175 at or above	1,000	175
25 "	999¼	200
15 "	998	215
30 "	997½	245
10 "	997	255
25 "	996½	280
25 "	996	305
20 "	995½	325
200 at or below	993	525

Source: Ahmed Zubeir Noordin, The Penang Tin Market, Proceedings of the Conference on Tin Consumption, Kuala Lumpur, 1974, Vol. IV, p. 49.

shown in Table 7.1, assuming that 270 tonnes of tin are offered for sale. The cut-off price in this case is M$996½ per pikul, those bidding at this price being rationed to the extent of the balance available from the sellers' offer.

The supply of concentrates coming forward from the mines to the smelters is influenced by miners' interpretation of market trends, but they do not usually hold accumulated stocks for long, which limits their ability to 'play on offers to influence the price'. Within limits, however, they can choose the right market day for selling their concentrates, an option which must be exercised within 28 days of the date of delivery of the concentrates to the smelter or its agent. As far as buyers are concerned, it could be argued that the secrecy with which bids are handled discourages any attempt by buyers to exert a collective influence on the price. Too unrealistic a price bid by a buyer would risk outbidding and hence no tin.

The Penang pricing system has been criticized, partly because of the secrecy with which it works, partly because of the lack of forward dealings. Market opinion seems to differ on the need for change, but the new Kuala Lumpur commodity exchange is expected to involve changes in the traditional Malaysian marketing system in the eighties. Whether the development of tin trading on the Kuala Lumpur exchange will seriously affect the LME remains to be seen.[5]

The LME is a very long-established system for the marketing of a number of non-ferrous metals, including tin. Normally only a small percentage of the total turnover on the LME represents actual metal transactions. In 1979, official cash and forward transactions, together with 'kerb' transactions (after the end of formal dealings), amounted to 257,980 tonnes, compared with Penang transactions in physical metal amounting to 57,745 tonnes. On the LME, physical dealings in tin metal were probably less than 15 per cent of total turnover. During the sixties and seventies there was a large increase in the LME turnover. Between 1965 and 1975 turnover increased threefold, and by a further 50 per cent in the next four years. It seems likely that part of the increase, at least, may have been due to speculation over currency uncertainties, since tin, as a high value non-ferrous metal, tends to attract speculative interest in conditions of inflation and currency instability.

Speculators make extensive use of the LME to profit from accurate forecasting of the movements of a fluctuating metal price without the obligation to take or make delivery of metal. Their participation helps to broaden the market and facilitate hedging or insurance operations by dealers or users, who want effective protection from price uncertainties. Many buyers on the Penang market hedge their purchases on the LME, which has remained a key factor in pricing and the only hedging institution for tin. The persistence of the LME as the main focus for price determination in spite of the absolute and relative decline of the UK in the commodity field must reflect its utility to both producers and consumers. However, it has often been the subject of criticism.

It is frequently alleged that the LME actually encourages price instability as a result of the volume of speculative activity which it naturally attracts. Even a normally sympathetic observer was recently led to comment:

> Future trading is a flourishing aspect of world trade, but a most disturbing feature became increasingly apparent in 1979 . . . There are extremely powerful agencies prepared to manipulate the free markets to further their own ends. We have seen this on the Comex silver and copper markets, and during June on the LME one particular company ruthlessly distorted the tin market.[6]

The evidence on which this criticism was based was a very marked divergence in June 1979 between price movements on the LME and those in New York and Penang. The LME price in June ranged from a daily low of £7,460 a tonne to £8,125 a tonne, a fluctuation of just under 10 per cent, the trough and peak being reached within a few days. The range on the Penang market was only M$10 a pikul, or 0.5 per cent, and on the New York market 23 cents, or under 3 per cent, namely from 700 to 723 cents a pound. In the four months April to

July 1979, the LME price rose by 15 per cent and fell by 18 per cent. The corresponding fluctuations in Penang were 4 per cent and 5 per cent, in New York 6 per cent either way.

The argument over speculation and the role of the LME as a price fixing agency has been going on for many years, not only in connection with tin. From his vantage-point as secretary to the International Tin Council, Fox has commented:

> the LME seems at all times to show undue complacence about the more speculative uses which have been made of the market. It may well be asked whether more authoritative or representative bodies, whether speaking for tin consumers or tin producing governments or for the ITC, might not put enough pressure on the Exchange to ensure that its internal affairs are so arranged as to minimise the impact of massive speculative dealings. It is certain that, while no action is taken, things will remain as they are, and the ability of the ITC to stabilise prices through the LME will always be limited.[7]

Critics maintain that many price movements on the LME, especially very short-term ones, are not justified by basic changes in supply-demand conditions. Prices in general on the LME, according to this train of thought, are much more volatile than they need be. On the other hand, it might be pointed out that prices were volatile during the post-war years when the LME was not in operation, namely from 1945 to 1953. There have also been periods during which the LME seems to have assessed better than producers or users the future movement of prices, as happened with copper in the first half of the sixties.

Some critics seem to favour producer pricing. At the 1974 tin conference, one speaker asked why consumers did not buy directly from producers, pointing out that with copper many consumers had direct annual contracts with the main producers, the basis being the LME price quotations, but at times copper producers have fixed their own contract prices to buyers.

Producer pricing is used in one form or another for nearly all metals, tin being the chief exception.[8] For consumers it has the advantage of a relatively stable price and possibly security of supply. As long as producers charge the same price to other consumers, there is no risk of being undercut. Producers would no doubt aim at an adquate long-run price, with gradual increases allowing for general inflation – in fact, a kind of producer-administered indexation. Against producer pricing is its relative inflexibility, its inability to respond sufficiently rapidly to changing market conditions. Producers may also make mistakes in price fixing, for example, confusing short-term fluctuations and long-term trends in demand, as copper producers seem to have done in the early sixties in an experiment with administered pricing. At that time the LME was a better guide to a changing market situation.

How widespread producer pricing is at the present time is uncertain. It works apparently with considerable success in the US copper industry, but only because there is a substantial degree of vertical integration in the US. Even in the US not all copper is subject to producer pricing, a sizeable proportion is fixed on prices closely linked to those of the LME. So far, there is little evidence of producer pricing in the world tin market, even with the state mining corporations. It is possible, however, that the existing trading system may gradually change in the direction of arrangements between producers and consumers. This is the view of a US consultancy group which suggests that a smaller percentage in the eighties will be handled by traders, with the result that 'consumers will have opportunities to make long-term arrangements with producers under more favourable conditions than in the past'.[9]

Notes

1. For a detailed account of the marketing system see C.A.J. Herkstroeter, 'Some Aspects of the Marketing of Tin', also Ahmad Zubeir Noordin, 'The Penang Tin Market', papers in the *Proceedings of the Fourth World Conference on Tin*, Kuala Lumpur, 1974, Vol. Four. This section draws heavily on these accounts of the marketing system.
2. Knorr, op. cit., p. 62. See also W.Y. Elliott and others, *International Control in the Non-Ferrous Metals*, Harvard Univ., New York, 1937, pp. 288–90.
3. For information on smelting costs the writer is indebted to Bernard C. Engel, Deputy Buffer Stock Manager.
4. *Tin International*, June, 1975.
5. The new Kuala Lumpur Commodity Exchange was opened in October, 1980, dealing initially with palm oil futures. The President of the States of Malaya Chamber of Commerce and chief executive of the Malaysian Mining Corporation stated that 'trading through the Penang Market will have to cease' when tin trading develops on the new Exchange. See report in *Tin International*, December 1980.
6. *Tin International*, January 1980. See also the account of and comments on the events of June 1979, in the same journal, July 1979.
7. Fox, *Tin: the Working of a Commodity Agreement*, p. 400.
8. See the RTZ memorandum to the House of Lords Select Committee on Commodity Prices, *Minutes of Evidence*, Vol. Three, 1977, p. 531.
9. *Tin International*, February 1981, reporting on a study of the prospects for tin by Emory Ayers Associates, New York, 'A Five-Year Outlook for Tin: Information for Executive Decisions'.

8 THE INTERNATIONAL TIN AGREEMENT AND PROBLEMS OF MARKET INTERVENTION

Origins of the ITA

Tin has been unique among minerals in being the subject of a fully-fledged intergovernmental commodity scheme since the 1930s. The first postwar agreement did not come into effect until 1956 and subsequent agreements then followed at five-yearly intervals. The fifth agreement, however, has been extended until 1982 because of the delay in completing negotiations for the sixth agreement. Over the last 25 years, a large number of developed and developing countries, with a variety of often conflicting interests, have maintained a continuous dialogue on commodity problems affecting the world tin market. It could almost be described as a microcosm of UNCTAD. Without changes in the basic principles, continuity has been maintained by successive agreements.*

The question naturally arises – why should tin, of all the internationally-traded minerals, have had an international agreement and, moreover, one which has lasted for so long in the varying conditions of the postwar world? All experience has proved that it is extremely difficult to organize international commodity agreements and to keep them operating successfully for any length of time. Since the Second World War there have been only five other primary commodities for which international agreements have been successfully negotiated.[1] None of these commodities has had such a long continuous history of international agreement. Despite the intensified pressure for commodity agreements during the seventies there has been little progress. The record of the ITA, therefore, seems even more unusual, set against such a background.

Intervention in the world tin market in one form or another goes

* The most detailed study of the long history of international tin agreements in the interwar years and up to 1973 is the authoritative book by William Fox, Secretary of the International Tin Council (1956-71), Tin: the Working of a *Commodity Agreement* (Mining Journal Books, London, 1974). The interwar experience of international tin control and the economics of the world tin market before 1944 were analysed by K.E. Knorr, *Tin under Control* (Stanford University, Calif., 1945). There is also a very useful account covering the period up to the mid-sixties, by Yip Yat Hoong, *The Development of the Tin Mining Industry in Malaya* (University of Malaya Press, Kuala Lumpur, 1969).

back a long way, in fact to the early twenties. The history of the inter-war intervention has been well documented. Only the salient features need be reviewed.

Instability on the tin market was very marked long before the First World War, but not outstandingly so compared with the markets for other non-ferrous metals. A serious surplus problem did not arise until after the war. The first intervention was the so-called Bandoeng Pool, an intergovernmental arrangement between the Federated Malay States and the Dutch East Indies to take surplus tin off the market in 1921 during a period of weak demand and low prices. The stocks acquired by the Pool were liquidated profitably by 1924 without disrupting the market, but there was a price for success, possibly realized only later. This is the view of K. Knorr, discussing the effects of the Bandoeng Pool on the world tin market both at the time of support and subsequently.

> In retrospect, government intervention in the form of the Bandeong Agreement had one profoundly disturbing effect on the future of the tin mining industry. The operations of the Pool contributed materially towards the building of surplus output capacity. The gradual liquidation of the stocks in 1923–24 disguised the fact that growing tin consumption was running ahead of the existing capacity to produce. In the absence of sales from the Pool stocks, tin prices would have risen more sharply at an earlier time and would, thus, have pointed to a growing disequilibrium between output capacity and capacity to consume. As it happened, it was not before 1925–26 that the tin market and tin producers generally realized the insufficiency of existing output capacity. Then tin prices soared to an exceedingly high level. The violence of this reaction, in turn, over-stimulated investment and thereby made for the creation of surplus production capacity. The Pool operations certainly did not cause the tin boom of the middle 1920s, but they contributed to its magnitude.[2]

During the thirties three successive international agreements came into effect aimed at supporting the market price, first by quotas, and from 1938 also with a buffer stock. Member countries had serious difficulties over the so-called 'standard tonnages' in regulating exports, in particular with those to be allocated to prospective new members. After the initial price collapse during the early part of the depression, which affected all primary commodities, the price of tin was successfully

Table 8.1: Selected Non-ferrous Metal Prices,[a] 1920–1939 (£/tonne)

	Tin	Copper	Lead	Zinc
1920	291.4	108.8	37.2	43.7
1921	162.8	73.9	22.4	25.4
1922	157.0	68.4	23.7	29.5
1923	199.3	71.4	26.7	32.5
1924	244.9	67.2	33.9	33.2
1925	257.0	66.0	35.9	36.0
1926	286.6	64.7	30.6	33.6
1927	284.5	61.3	23.8	28.1
1928	223.6	68.4	20.7	24.9
1929	200.7	84.1	22.9	24.0
1930	139.7	61.2	17.8	16.3
1931	116.6	42.0	12.8	12.0
1932	133.8	35.8	11.7	13.3
1933	191.5	36.1	11.5	15.4
1934	226.7	33.0	10.8	13.4
1935	225.7	35.1	14.0	13.9
1936	204.6	42.2	17.3	14.7
1937	242.3	59.1	23.0	21.9
1938	189.6	45.1	15.0	13.8
1939	226.3	47.9	15.1	14.0

Note: a. Average annual LME prices.

Source: C.J. Schmitz, *World Non-Ferrous Metal Production and Prices 1700-1976*, Cass, London, 1979.

raised to a relatively high level for the time. Bearing in mind the generally low price levels on commodity markets in the thirties, it can be seen from Table 8.1 that tin did well compared with other non-ferrous metals.

It could be argued, therefore, that the market intervention was successful, although there was a price in terms of output foregone, and the gains were by no means equally shared. Member countries, however, seemed to consider that the balance of gains and losses was positive. This was one reason why the idea of an international agreement after the Second World War appealed to most producers. Secondly, the experience of operating a control scheme over a considerable period showed its practicality. It was possible to reach workable compromises and to avoid 'chiselling' by disgruntled producers. Thirdly, producers did not see how a situation of actual or potential surplus could be handled in the postwar period, after the recovery stage, other than by an international agreement relying on some combination of export controls and a buffer stock.

From the point of view of the governments of some consuming countries and of major industrial users, especially in the US, the prewar producers' organization had the demerits of a cartel. While recognizing the deficiencies of a free market under certain conditions, which made some form of intervention possibly desirable, the consuming countries believed that their interests needed effective representation. The US was particularly interested in curbing international cartels affecting industrial materials and there had been much discussion in a number of countries during the thirties about freedom of access to internationally-traded materials. Since the US was the dominant economy in the forties, it is not surprising that the postwar international system was intended to favour joint consumer-producer arrangements, in so far as any kind of international scheme for primary commodities was acceptable to the US.

When the question arose of future international arrangements for tin to replace the prewar scheme, negotiations involved both producing and consuming country governments under the auspices of the UN. The negotiations were long-drawn-out, difficult, and at times acrimonious, before eventually a formal joint producer-consumer agreement was settled between governments. Between 1948 and 1953 the Tin Study Group put forward a series of draft agreements as attempts were made to reach a compromise acceptable to the various interests.[3] Although an agreement was reached, the controversial issues of these years of negotiations have continued to be argued about among member countries and threatened the survival of the agreement in the 1980s.

In the negotiations, the UK, Belgium and the Netherlands were more sympathetic than the US to the producers' case, being heavily involved politically with producing countries in the fifties, as they had been in the interwar years. A serious dispute between producing countries and the US over the prices during the Korean War complicated negotiations. Between the Korean commodity boom and 1953, however, the US attitude softened, possibly because the State Department had to take into account the political repercussions of the hostility between the US and the developing countries. Nevertheless, the US did not sign the 1953 agreement, although it adopted a much more conciliatory attitude.

Since 1956, when the 1953 agreement became effective, there have been five sets of negotiations. The latest, to replace the fifth agreement due to expire in June 1982, proved exceptionally difficult. There has been an impressive list of objectives, much of it, in the opinion of W. Fox, largely 'cosmetic', since the Tin Council had neither the resources nor the authority to do much about most of them. The main objective throughout the history of the five agreements has been to prevent excessive price fluctuations. Other, longer-term, objectives have been to prevent serious imbalances between supply and demand, to increase tin export earnings, to secure adequate supplies for consumers and to achieve a dynamic and rising rate of production. As far as prices are concerned, the agreements refer to fair prices for consumers and a reasonable return for producers, but, understandably, there is no definition of what is meant by 'fair' or 'reasonable'.

Although the objectives indicate the efforts to achieve a compromise, Fox points to a basic asymmetry in the treatment of surpluses and shortages. With a surplus, beyond the use of the buffer stock, the section of the agreement dealing with export control is strict and compulsory. In contrast, 'the parallel article on action in the event of a shortage has only two clauses, and the compulsion there on the ITC is merely to enquire, to recommend and to observe'. In Fox's view, 'a real balance of interests and obligations would have required a change in the basic lines of the agreement which there was little likelihood would ever take place'.[4] The problems arising from such an attempt were shown clearly by the negotiations in 1980 for the sixth agreement.[5]

Organization of the ITA

The postwar agreements have always had a strong producer representation, with only a small part of non-communist world production outside their scope. The chief non-communist non-member is Brazil, with less than 4 per cent of world production and a negligible share of world exports. The only serious, if intermittent, problem of an outside producer has come from China. The Soviet Union is an important producer, but has been a net importer, except for a short period in 1957–8 when it was a seller of tin, formerly imported from China, on a weak market. The problem was eventually tackled by an embargo on imports of tin from the Soviet Union by several consuming members, and an agreement by the Soviet Union to curb its sales, which ceased to be a problem in 1959. The Soviet Union joined the fourth agreement as a consuming member, but used as a basis for voting its imports rather than its consumption, which, like its production, remains unknown. As a consuming member it is not required to supply the Council with information about its production. The only producer loss has been Rwanda, included with the Belgian Congo in membership of the first agreement.

On the consuming side there was a major absentee until 1976, when the US at last became a member. Between 1962 and 1976 Japan had the largest single consumer vote. Since 1976, the US and West Germany have had enough votes jointly to veto decisions requiring a two-thirds majority of both consumer and producer votes on key issues, notably determining the buffer stock price range. The UK, Belgium and the Netherlands were members from the beginning. Each still had political links with producing countries in the fifties and their nationals owned substantial tin mining assets. France was also a founder consuming member, and a country with a reputation for sympathizing strongly with international commodity control. The former secretary to the Council has stated:

> France had perhaps the clearest philosophy on commodity agreements and the developing countries. The government of France aimed in the Tin Council at being an essential link between the industrial and the developing countries, an essential link which would bring France status and prestige and might help developing countries to act within the limits of economic reason.[6]

West Germany and the US seem to have been the most intractable among the consumers, the most resistant to proposals for increases in the buffer stock price range, the most critical of some policies in the developing producing countries. Although the Council is naturally secretive on its internal debates, some information filters out. Fox has stated that 'in general, the consumers did not generally show the same community of interest as did producers . . . the Netherlands, Belgium and Spain could normally be expected to vote with producers on major points of importance; the UK tended to hedge; France poured oil on troubled waters; Japan, and to a lesser extent, Canada, stood firm in resisting proposals for increases in the price scale of the agreements'.[7] Little information seems to be available on the attitude of the centrally-planned countries to important issues, even after the Soviet Union became a member in 1971. All the other East European members of the COMECON, except East Germany, are also members, with a combined voting power in mid-1980 roughly equal to that of the UK, France and Italy.

The Buffer Stock, Prices and Export Control

Sharp differences of opinion over the buffer stock arose during the negotiations for the first postwar agreement. The producing countries wanted the limit to be 15,000 long tons at a time when consumption averaged 135,000 tons. The US, which was a participant in discussions of the various drafts before final agreement, proposed a buffer stock of 35,000 tons, but neither the US nor any other consumer at that time was willing to contribute to its financing.

There were several reasons for producer country opposition to a larger buffer stock. With a high opportunity cost of capital, they naturally did not want to put much money in a buffer stock. However, as Fox points out, 'the contributions to the buffer stock were, except in the nationalised industries in Bolivia and Indonesia, raised by the local governments from the private miners; this meant that the British-owned mines in Malaya and Nigeria and the Belgian-owned mines in the Congo for many years helped in the capital contribution'.[8] A more important reason was that the producing countries did not believe that a buffer stock would give a guarantee of success in defending the market price. They also disliked the idea of a large buffer stock hanging over the market. And, finally, they believed then, and have always argued, that it was inequitable that the entire burden of financing the buffer stock should fall on the producing countries.

In the first agreement, the basic contributions of the producing countries amounted to the equivalent of 25,000 long tons of tin, representing about two months' consumption at the time. Subsequently the basic contributions have financed a smaller amount, but provision has been made to supplement the basic contributions by borrowing on the security of warehouse warrants for tin already acquired for the stock. An overdraft facility during the third agreement gave an additional £10 million. No use was made of the borrowing power until 1976, when Article 24a of the fourth agreement was used to borrow £26.1 million out of the £36.5 million standby credits. This enabled the purchase of 23,000 tonnes, instead of the 12,000 tonnes that was all that the members' contribution could buy. Even so, the amount that could be taken off the market then was much less than in 1957–8, when a supplementary facility was arranged with a group of banks. Since 1976, the financial position of the buffer stock has improved. Under the fifth agreement, the equivalent of producer contributions was 20,000 tonnes, plus voluntary consumer contributions which should have matched that amount. With borrowing on the strength of warehouse warrants it became possible, in principle, to buy up to 90,000 tonnes, according to the Tin Council chairman.

Consumer opposition to compulsory contributions, as applied to producing members, has been a persistent grievance to the latter, particularly as consuming members vote on the use of the buffer stock and the relevant price range. Opposition could hardly have been justified on the cost of contributing. Presumably there was the risk of creating a precedent, although the absence of other agreements did not threaten consuming countries with a flood of demands for buffer stock money. Compared with the large sums of money spent on foreign aid, often with dubious results, the cost of financing a larger tin buffer stock would have been minute.

Moreover, the financial outcome of the tin buffer stock seems to have been satisfactory, although in the absence of exact cash flow information about its operations, it is not possible to determine precisely the profitability of the buffer stock by means of the standard method, namely, the discounted cash flow method. Lacking this information, C.P. Hallwood has attempted an assessment of its profitability from 1957 to 1977 on various assumptions about market intervention and the opportunity cost of capital.[9] Using discount rates of 8 per cent, 10 per cent and 12 per cent for the 1970–7 period and 8 per cent for the previous years, Hallwood calculates that, in general, the tin buffer stock was self financing. The financial position as revealed by the Tin Council shows that in 1961, on the conclusion of the first agreement, there was a profit of £4.4 million on contributions of £16.9 million, representing 5.2 per cent at an average annual rate. £1.4 million came from investment income, the rest from the profit on selling at a higher price than the purchase price. Under the second agreement, a £14.6 million contribution led to a surplus of £4.8 million, almost all of which came from money market operations by the buffer stock manager. On average, the buffer stock operations earned an annual undiscounted profit rate of about 8 per cent on subscriptions up to 1976, partly from investment income and partly from tin transactions over a long period of rising prices. All buffer stocks in tin, both before and after the Second World War, have made a profit for their participants, but whether this matched the opportunity cost of capital in developing countries is uncertain. At least it is important that they did not run at a loss. However, the surpluses which were realized from the postwar buffer stocks have apparently not moderated the disagreements over the use of the buffer stock which have persisted for many years.

Normally the buffer stock manager is required to follow the rules laid down in the agreements; no intervention when the market price is in the middle sector of a three-part range; obligatory sales at the upper limit, obligatory purchases at the lower limit; discretionary selling or buying in the upper sectors respectively, in order to check too rapid a movement in either direction. On several occasions, intervention has been permitted within the middle sector, but the buffer stock manager has not been allowed to lead the market, nor to try to keep the market price within that sector. On each occasion export control was also in force. Export control, in fact, was used well before the buffer stock's normal buying power was exhausted. Consumers would have preferred less use of export control. The US, joining only the fifth agreement, was strongly opposed to export control. The Canadian view, representing a country with a very large mineral export trade, apparently preferred export control only as a last resort, when the market price fell below the floor price. Fox has stated that Bolivia thought that 'the objective of the agreement was to obtain an actual market price around the mid-point between floor and ceiling prices'.[10] Both Bolivia and Indonesia, according to Fox, wanted buffer stock buying in the middle sector. As generally the highest cost producer, Bolivia has always wanted as high a price as possible, and has usually led the push for a higher price range. Raising the floor price to match Bolivia's costs usually meant raising the ceiling price, since there was little scope for narrowing the range.

The choice of price limits for buffer stock operations has naturally been the crucial issue and the chief source of conflict throughout the history of the postwar tin agreements. Several problems have been raised: the width of the price range, the relationship between the buffer stock limits and the shifting market price, the provisions for altering the price range, the width of the three sectors. Up to March 1957, the range was at its widest, a floor of £640 a tonne and a ceiling of £880, 37.5 per cent measured in relation to the floor. Since then the range has normally been not less than 20 per cent and not more than 27.3 per cent. For a short time in 1963–4 it fell as low as 17.6 per cent. In July 1979, however, a sharp increase in the ceiling price raised the range to 30 per cent, a recognition of the fact that for some time the market price was above the ceiling. Even so, the ceiling had to be raised again in March 1980, after a further rise in the market price. The floor price was adjusted to maintain the 30 per cent range.

Although generally speaking the width of the range appears to have been a reasonable compromise between an impracticably narrow one and a wide range needing less intervention but giving less stability, there have been a number of problems in its operation. The producers aimed at 'remunerative' prices, with a high floor price, and were not greatly interested in defending strictly a ceiling price. Consumers were more interested in a relatively stable price and a lower ceiling price than the producers wanted. There was no agreement on the meaning of a 'remunerative' price. For governments of developing countries, a 'remunerative' price seems to be one which not only covers the commercial costs and profit aims of miners, but also leaves a substantial margin for taxation, including export duties. The government, in short, wants to maximize its income from the mining industry. Some consuming countries have opposed changes in the buffer stock price range because they believed that the tax policy of producing countries prevented higher prices stimulating higher production. For example, in 1979 there was strong opposition to a higher buffer stock price from the US, even although this would have simply meant recognizing the unreality of the existing ceiling, when the market price was either above or close to it.

The American attitude was that the market price was artificially high and not 'a true equilibrium price set by market forces'.[11] The market, according to this view, was distorted by government action which prevented supply adjustments to high prices. In effect, the US was criticizing the right of producing countries to pursue a particular domestic policy which they believed in their national interests, and rejecting the idea that the ITA should be used to give its rubber stamp approval to a policy which interfered with normal market forces.

In so far as the buffer stock range has been kept too low for a number of periods, the Tin Council has been committed to the defence of an unrealistic ceiling with inadequate resources. Since there was an overriding obligation to sell any tin in the buffer stock at the ceiling price on a rising market, speculators could not fail to make a profit by clearing out the stock, which they did on several occasions. On the other hand, the unrealistic range made it easy to maintain the floor price, especially as the buffer stock funds could be backed up by effective export controls.

To some extent, agreement on the price range was hampered by a lack of reliable information on costs of production throughout the industry. In March 1977, an important innovation was made in setting up an Economic and Price Review Panel, which was intended to help bring about a more objective and scientific approach to the assessment of mining costs and hence a more realistic price range. By this time, of course, the whole international commodity debate had been affected by the great changes in the oil industry and by the demand of developing countries in general for a new international economic order.

The Bolivian position, representing the extreme producer view, was that Bolivian costs should be covered. Further there was a general view among the developing producing countries that the buffer stock price changes should be governed by the principle of indexation, which would take account of inflation in the prices of manufactures and other elements in the cost structure of the mining industry. It was argued by the Bolivians that indexation would avoid 'discussions which foster antagonisms and lead to polarisation reflecting politican stances totally foreign to the technical considerations towards which the Council ought to address itself'.[12]

The Review Panel certainly brought more factual evidence to bear on the question of the buffer stock price range, but it did not still the arguments on the true costs of production, nor did it bring about agreement on the disputed issue of indexation.[13] Indexation has not been accepted by industrial countries either for tin or for other important industrial materials. Opposition from consumers to raising the floor price, and hence the ceiling price usually, has continued for the usual reasons: scepticism about the pressure of costs, an implicit dislike of indexation, criticism of tax policy towards mining in most producing countries, and, especially from the large steel firms, the effect of higher tin prices on the cost of tinplate.

Whether indexation would be as damaging as is usually maintained is debatable. In a study of the tin agreement, C. Gilbert has argued that it was a mistake not to make automatic adjustment of the floor and ceiling prices in line with inflation, irrespective of the currency in which the price was quoted. In his opinion, 'it is in the context of buffer stock agreements that the widely-voiced sentiments in favour of indexation of primary product prices can be most simply given practical content'.[14] Gilbert suggests that indexation of the support price was justified, and illustrates what it would have meant for the first four agreements by updating the 1956 price level by the increase in the UK wholesale price index, plus 3 per cent per annum. The resulting prices up to 1971 were as follows:

	Actual initial floor price	Indexed basis initial price
	£ per tonne	
First agreement, 1956	630	630
Second agreement, 1961	719	792
Third agreement, 1966	1,083	1,041
Fourth agreement, 1971	1,350	1,536

Gilbert's calculations imply that producers would have obtained a higher initial floor price with indexation according to his formula under the second and fourth agreements.[15] Under the fourth agreement the initial indexed basis price would have been 14 per cent above the negotiated floor price. In June 1975, according to Gilbert, the indexed floor would have been £2,955 compared with £2,739, the renegotiated floor price. The whole range would also have been moved up, assuming that there was no difference in the width of the range under indexation.

Whatever the merits of indexation in principle, it would be a mistake to assume that it would be an easy solution to the price problem. Certainly in a world of fluctuating exchange rates and of widely varying rates of inflation in the industrialized countries, the choice of a price index for indexation becomes extremely complicated. It must be accepted that there would be difficulties in determining the influence of inflation on the long-term equilibrium (nominal) price. Further, irrespective of attempts at introducing a more scientific method of price determination, the target price range employed in practice by a buffer stock agency is likely to be a negotiated one between consuming and producing countries. Governments would not hand over the vitally important and controversial issue of the price range to an automatic formula. Hence, adjusting the target price for changes in some world price index would simply add another arbitrary element to the process of determining the price range and this might move the target price away from or toward the appropriate long-run equilibrium price.[16]

The US Strategic Stockpile

For 40 years the US stockpile and stockpile policy have been a major factor in the world tin market.[17] A stockpiling programme was inaugurated in 1939, and in the following year the US became a large buyer of tin for the stockpile. It was apparent that heavy buying by the Metal Reserve Company would disturb the market unless tin production increased at the same time. The American government entered into direct negotiations with the International Tin Council, which agreed to raise export quotas from 80 per cent during the second quarter of 1940 to 130 per cent for the year as a whole. The government offered to take up all surplus tin, up to a total of 75,000 tons, at a minimum price of 50 cents a pound, all purchases being made during the 12 months ending 30 June 1941. Imported concentrates were to be smelted at a new government-owned smelter in Texas. Even at this time it was obvious that a very large stockpile would be accumulated, since a sum of US$150 million was allocated for tin purchases, equivalent at the floor price to about 150,000 tons.

There were different views on the question of the stockpile. The official aim was a large reserve of tin to avoid the disruption which might be brought about by the interruption of supplies from southeast Asia if the war spread to that area. Some industrial consumers in America saw in the stockpile 'the establishment of an objective long-cherished, namely a potent consumers' pool'.[18] From the beginning, producers were uneasy about the long-run implications of a huge stockpile. For this reason a clause was included in the contract between the ITC and the Metal Reserve Company, which specified a particular procedure to govern the eventual liquidation of the stock. According to the agreement, apart from a national emergency, the US government committed itself to dispose of the stock 'only upon 3 months' notice to the ITC, and at an annual rate not exceeding 5 per cent of the aggregate stock or a maximum of 5,000 tons'. Nevertheless, as Knorr points out, 'unless this government reserve should be exhausted in the course of a national emergency, it appeared at the time that the stock constituted some kind of potential consumers' pool of a truly impressive magnitude'.[19] After the war, unsettled political conditions in the main producing countries and the cooling of relations with the Soviet Union led to continued buying for the stockpile. The target was a stockpile which would be adequate for a four-year war with, presumably, the communist bloc. Unfortunately, it seems that stockpile buying got out of hand. By the time it was realized that a very large amount had been accumulated, enough tin was in the stockpile to meet six years' peacetime consumption at the rate prevailing in the mid-fifties. This must have been far larger than had ever been contemplated. The stockpile was more than twice as large as non-communist world production in the best year of the fifties. Of all the numerous commodities acquired for the strategic stockpile, none had been accumulated on such a lavish scale, none caused more trouble during both the build-up and the subsequent erratic, and still incomplete, run-down.

Within two years of the end of the war, the recovery of tin production exceeded commercial demand. This remained the position until the late fifties. During this postwar period, therefore, American stockpile purchases played a major role in the market. In fact, these purchases were a key factor in the rapid expansion of production, which could not possibly have been boosted to such an extent by the growth of commercial demand. Once the industry had recovered, as early as 1948-9, the continuation of American stockpiling helped the general economic recovery of several developing countries, and thereby contributed to their political stabilization, an important strategic consideration at the time, but it sowed the seeds of future trouble, and even at the time there were disadvantages.

During the panic buying at the time of the Korean commodity boom, the claims of the stockpile undoubtedly contributed to the phenomenal rise in prices. As the price rose, US dissatisfaction with the system for buying intensified, and charges of exploitation and 'price gouging' were levelled at the producing countries. The upsurge of criticism in Congress brought a new bitterness into relations between the developing countries and the US. In retaliation for what it considered unjustifiable prices, the US stopped buying for the stockpile. Athough there was only a short break before buying was resumed, the acrimonious dispute left a legacy of mutual suspicion.

During the fifties commercial buying by countries other than the US and the UK gradually increased, but up to the 1957-8 recession their purchases were still not enough to clear the market on their own. As late as 1957 production was about 8 per cent in excess of consumption.

Stockpile Disposal and the ITA

From the ending of stockpile buying until the first commercial sales by the General Services Administration, the US agency responsible for the stockpile, there was an uneasy interval of about six years. The US was not the first country to sell surplus tin. The UK sold 5,000 tons in 1959-60 when output was subject to export controls, and some sales were also made by Canada and Italy. The total amount available in these stockpiles was minute compared with the officially published figure of 349,000 tons in the US strategic stockpile at the end of 1961.

Once it became apparent that there was a huge amount of tin which was no longer necessary for strategic purposes, and that there would inevitably be pressure to get rid of it, the producers saw their future prospects dimmed by the unfair use of American bargaining power. It was argued that an earlier US government had given a binding commitment not to release surplus tin except in an 'emergency', which was understood by producers to mean virtually a state of war. Sales would otherwise be a breach of faith, since the reference to an 'emergency' had been explicitly included in the official minutes of meetings between the US government and the government of the Dutch East Indies, which had supplied about half the tin.

It is doubtful whether the government after the war accepted the producers' interpretation of the earlier agreement. Moreover, the commitment, however interpreted, had been given to a colonial government, which had since been replaced by an independent Indonesia. But Indonesia, like other developing countries, was heavily dependent on commodity exports, and the US was supposed to be committed to aiding their development. It seemed both illogical and unjust to the producing countries that the US should contemplate a huge disposal programme. Fox takes a very critical view of US policy on stockpiles. The accumulation of the stockpile he regards as 'irresponsible', in that it went well beyond the objectives, and had encouraged too rapid a growth of production in postwar years. In his opinion, if US purchases had tailed off gradually to a much lower total, it would have been easier for the industry to adapt itself to normal commercial demand after the war.[20]

By 1960 it was evident that with a lack of buoyancy in American consumption, the stockpile alone could take care of at least six years' normal demand. Fortunately for the disposal programme, the shortfall of production in the early sixties led the Tin Council to turn to US sales as the only way of preventing a runaway increase in the market price. In 1964 and 1965 these sales were equal to 22 per cent and 15 per cent of current production. Buffer stock supplies had been finally absorbed by speculators in 1961. With the ending of export control in 1961, and serious difficulties in several countries, only the stockpile was available. In 1964 the US initiated a long-range disposal programme for the tin which was declared surplus to the strategic stockpile, some 148,000 tons. It was proposed to sell this amount over a period of 6–8 years. The average annual rate of 18,500 tons would have been over three times the rate proposed in the 1940 agreement between the government and the Tin Council. The disposal rate proved overoptimistic. Only two-thirds was sold in eight years. No commercial sales were made in 1969-72, but other disposals amounted to over 7,000 long tons. The programme announced in 1964 was not completed until 1978, by which time the disposal of another large amount was under consideration.

The GSA generally sold at the ceiling price, partly out of consideration for the developing countries, partly to maximize the return. Nevertheless, the disposal programme was the cause of serious discord between the US and producers. One reason was the uncertainty over disposals, since the 1964 programme was subject to many changes. The US government could not decide on a consistent policy for the stockpile. However, there was only one occasion on which the GSA sold tin when the Tin Council was using the buffer stock to support the flow price. In 1966 the GSA sold a large amount of tin within a period of three weeks. In effect, the buffer stock took up the surplus.

At this stage, the strong complaints of producers were reinforced by the US State Department's anxiety over political repercussions. There was inconsistency between a policy of converting surplus tin into cash, which could be only of very modest budgetary or balance of payments importance, and the permanent policy of supporting developing countries. Political issues loomed large in 1966-7: control of raw materials in Zaire, access to raw materials in Indonesia, political instability in Bolivia, the role of Thailand's airfields in the Vietnam war. The State Department took the view that 'in such an atmosphere, the few million dollars immediately involved in pressing on with a tin disposal programme in very sensitive international areas were of no importance'.[21] In October 1966, an informal agreement was reached between the GSA and the ITC, whereby the GSA promised that it would not be a seller when surplus tin was being taken off the market by the buffer stock manager. This concession did not prevent the stockpile remaining a cause of friction between the producers and the US.

The mere existence of a large stock is a disturbing influence in the market once it is no longer believed to be in absolutely firm hands with a cast-iron guarantee against disposal. Clearly, the US has had no settled policy, since there were no less than 13 changes in the amount to be reserved for strategic purposes between 1944 and 1977 (see Table 8.2). Further, the legislative processes of deciding on the size of the surplus, the timing and rate of disposal, have always been long drawn out, with lengthy and confused debates in Congress. The precise

Table 8.2: Changes in US Strategic Stockpile Targets, 1944–1976 (tons)

November 1944	210,000
December 1949	260,000
January 1950	285,000
November 1950	350,000
June 1951	245,000
September 1954	308,000
March 1955	341,000
June 1958	198,000
July 1960	185,000
May 1963	200,000
March 1969	232,000
April 1973	40,000
October 1976	32,000 (total holdings 202,842)

Source: *Tin International*, December 1976.

role of the stockpile has never been determined. How much was needed for a military emergency? It could hardly be for a war with the communist bloc, for if that were the case, it might be held for ever, at least if one discounts a nuclear conflict between the super-powers. In the event of such a nuclear war, there would be less need for tin, since there would presumably be far fewer people to consume it. Was a large stockpile necessary as a safeguard against a temporary interruption of supplies from politically unreliable developing countries? It could hardly require a stock which by the mid-seventies was still equal to more than three years' normal demand. What was supposed to be the relation between the official stock and the private user?

There is some evidence that private enterprise reduced stocks once it was obvious that the government did not intend to hold on to the stockpile indefinitely. An ITC Working Party maintained in 1963 that US stockholding habits had changed as a result of the first sales by the GSA. More recently, in evidence to a Congressional Sub-Committe on the possible effects of the stockpile, Dr. G.W. Smith has argued that 'new investment in tin production and in private speculative stocks were no doubt reduced below what they would otherwise have been'.[22] As Smith pointed out to the committee, however, the delays in reaching decisions and giving effect to them meant that when the shortage developed at the end of 1976 and in 1977, the GSA was not yet authorized to sell tin. It was not until October 1977 that the government at last made up its mind about the 32,499 ton target for the strategic stockpile, and it still had to get Congressional authorization for selling the excess tin. In fact, the GSA was not ready to offer tin until mid-1980, by which time, although the market price was still relatively high, production had reached a postwar peak, and only net sales to the centrally planned countries prevented a surplus problem.[23] Since the Soviet Union is a net importer, GSA sales would indirectly increase the supply upon which the Soviet Union could draw, a strange result of a policy which was originally intended to protect the US against a foreign threat to an essential industrial material, of which the US was totally deficient.

The accumulation of stockpiled tin gave the US the equivalent of a major tin mine, in which the costs of extraction and smelting had already been covered in the past, leaving only the current holding charges. It could be argued, therefore, that it should be treated as a national asset to be kept as a reserve indefinitely, or, if it was politically impossible to insulate the stockpile completely, then a national long-term disposal programme should have been worked out by the US government. One proposal was put forward tentatively by Fox on the basis of his long experience of negotiations between the Tin Council and GSA. Fox suggested that the surplus (i.e. the tin not specifically required for strategic purposes, however defined) should be treated

> as a tin mine, with a long-term working life, that this mine should be given an annual tonnage for disposal, that this annual tonnage should be left unchanged for a long period (perhaps even for the whole life of the mine) and that this should be made known to the world; that the arrangements for the world tin price should be left entirely to the ITC, and that there should be no imposition by the ITC of a minimum price below which no stockpile sales should take place.[24]

He also suggested that there might be a fixed annual tonnage of 5,000 tonnes (this figure appeared in the 1940 agreement between the ITC and the US), so that the full amount of the stockpile surplus should be spread over at least 40 years, taking the last of the stockpile well into next century.

The suggestion met with no response. It is difficult to imagine the US Congress accepting such limitations on American freedom of action. There must also be scepticism about the proposal that the surplus should be transferred to the ITC as a buffer stock. It has been difficult enough to win approval for a small transfer as the American contribution to the buffer stock under the fifth agreement.[25] As long as the US has a large stock, it has an implicit bargaining tool which could be used against a tin cartel of which Congress disapproved. Yet in assessing American interests, some account should be taken of the possible effect of the stockpile on investment in tin mining. Although it is impossible to prove, there are grounds for believing that uncertainty has had an effect. Admittedly, there have been other important deterrents, such as the land problem, taxation, political uncertainties and opposition to foreign participation in mining, but a large uncertainly held stock must be a market factor.

The ITA: Success or Failure?

The primary objective of the ITA has been to prevent excessive fluctuations in the price of tin, without supplanting market forces. Market forces have been allowed free play within the middle section of the price range. It was intended that buffer stock operations would exercise a braking effect if the market price moved too rapidly within the upper and lower sections. Export quotas were used at times to prevent a significant imbalance between production and demand. A judgement on the agreement rests heavily on its success in moderating price fluctuations, and in maintaining satisfactory floor and ceiling prices. It can also be judged according to its success in encouraging the long-term growth of production, ensuring adequate supplies for consumers, and raising the export earnings of producing developing countries.

The floor price was breached only for a matter of days in September 1958. The top of the range, on the other hand, has been exceeded for months or even years at a time, as happened particularly in 1964–6 and 1977–80. For 50 months out of 222 between 1956 and the end of 1974 the average monthly price exceeded the relevant buffer stock celing. The longest period in which the market price fluctuated in the neighbourhood of the middle sector of the three-part price range was from the middle of 1966 to the middle of 1973, and even in those years it was often in the upper sector of the range.

It is clear from the record that the agreement has worked only asymmetrically. It could be argued, however, that this outcome was the inevitable result of inappropriate buffer stock price ranges and the failure to adjust the range in line with basic market trends. If the price range became unrealistic, and the Council lacked the financial resources for a large enough buffer stock, market forces would eventually force the price into and beyond the upper sector. Since consumers generally resisted a revision of the price range in such a situation, the Council always lagged behind the market.

Once the buffer stock was exhausted, the only source of tin in the short and medium terms was the American stockpile. In so far as the stockpile was used to check a runaway price increase, it has been suggested by one critic that 'the interaction of the ITA and GSA produced a loose, informal, and unique commodity agreement in which the GSA had primary responsibility for defending fairly high and unknown ceiling prices, whereas the ITA was engaged mainly in defending fairly low but known floor prices'. In this writer's view, 'the effective band between ceiling and floor prices was quite wide, perhaps a ± 20 to 25 per cent variation around long-term trends, which is greater than usually envisaged for commodity agreements'.[26]

Different assessments have been made of the price stabilizing effects of the agreement. K. Backman found that price instability was much lower in the postwar period than in the interwar years, but noted that demand was more stable in the former period, which qualified any conclusion about the effectiveness of the agreement. [27] Her study used several measures: year-to-year average price fluctuations for periods with and without an agreement; within-year fluctuations; price fluctuations in other non-ferrous metal markets. Over the period 1956 to 1971, percentage annual deviations from the trend (using a five-year moving average) were lower than in the years 1923 to 1937. The contrast was quite marked, except for the years 1963 to 1965 when the average annual deviation shot up to 13 per cent compared with 3 per cent from 1957 to 1962, and 5 per cent from 1966 to 1971. Average within-year fluctuations around the average yearly price also showed a marked reduction compared with those in the twenties and thirties. The postwar period covered by the second measure included six years before the agreements came into operation. There was no significant difference between fluctuations in 1950–6 and 1957–71, but according to Backman, there was a statistically significant difference (at the 5 per cent level) between price fluctuations in 1950–6 and in the years when the buffer stock was used between 1957 and 1971.[28]

Backman's comparison with other non-ferrous metals showed that tin had a better record in respect of price stability up to 1971. Her conclusion is supported in a more recent study of primary commodities by I.S. Chadha, looking at the average deviation from trend of a number of non-ferrous metal prices from 1960 to 1975, using UNCTAD material shown in Table 8.3. It is also noticeable that in the more depressed conditions of the late seventies, the price of tin held

Table 8.3: Price Fluctuations in Leading Non-ferrous Metal Markets (average deviation from trend)

	Monthly prices			Annual prices	
	1960-4	1964-9	1970-5	1964-9	1970-5
Zinc	–	7.9	49.2	4.7	33.6
Tungsten concentrates	22.5	15.5	33.5	12.2	28.9
Copper	13.3	20.6	31.9	12.9	23.6
Lead	17.3	16.1	25.0	10.7	19.9
Tin	8.6	9.6	21.3	5.8	18.4
Aluminium	–	2.1	11.6	1.9	9.2

Source: I.S. Chadha, 'The North-South Negotiating Process in the Field of Commodities', in A. Sengupta (ed.), *Commodities, Finance and Trade: Issues in North-South Negotiations*, Frances Pinter, London, 1980.

up better than other prices, and produced remarkably high price ratios with other metals. The buffer stock ceiling, however, was breached, and the buffer stock was without tin for several years. The relatively high prices of the late seventies may have been the result of the failure of an earlier period of high prices to boost investment, as it did with other metals, and as it might well have done with tin also if supply conditions had been the same as in the past and there had been no threat from the stockpile.

Where the agreement has been an unqualified success since the end of 1958 has been the maintenance of the buffer stock price floor. There have naturally been vigorous complaints from producers that the floor price has often been too low and adjusted too tardily as a result of the reluctance of consumers to agree to a revision of the range at certain times. Nevertheless, a guaranteed floor price, even if lower than producers wanted, seems to have been useful, at least to the smaller producers, as the Chinese Mining Association in Malaysia has stressed.[29] The Association was in no doubt about the value of a guaranteed minimum price in helping its members to obtain local bank credit, upon which small gravel pump mines were heavily dependent. The view of the Association is supported by Gilbert, who goes so far as to say that 'it is the maintenance of adequate price levels in situations of prolonged and severe excess supply which is the most important function of commodity agreements, not price stabilization over the cycle'.[30] Moreover, in Gilbert's view, an adequate floor price in modern conditions would have to take account of inflation, which obviously implies an element of indexation in the setting of the buffer stock price range.

Fox also accepts that the floor price is an important target, but, writing in 1974, he was somewhat sceptical about its practical value in helping producers to raise new capital. He commented: 'It does not seem to have attracted the general flow of new capital that might have been expected, as there are no signs that the other non-ferrous metals, without the advantage of a floor price, have found any difficulty in raising capital'.[31] This judgement would seem to be applicable more to foreign-owned companies and to foreign capital than to local miners. Further, there were other factors which affected, in particular, the more capital-intensive projects: difficulties in reaching satisfactory arrangements with host governments, covering such things as taxation, leases and local participation, and possibly the US stockpile.

It is apparent that there are widely different views on the achievements of the agreement. Even the most sympathetic gives only a qualified approval. Possibly the least favourable judgement was made in 1976 by Smith and Schink on the basis of the Wharton model of the world tin economy.

> The Tin Agreement has only marginally reduced the instability of prices and producer incomes. Of far greater importance in this respect have been US government stockpile transactions made outside the Tin Agreement. The Tin Agreement endured while other agreements have failed, in part because it lacked effective power, in the face of the US strategic stockpile, to make critical price decisions which otherwise would have intensified producer-consumer conflicts. If the Tin Agreement had been designed from the beginning as an effective market stabiliser along the lines currently envisaged for other products, there is a good chance that it would have fallen apart.[32]

The Smith–Schink study envisaged a producer-consumer agreement intended to stabilize the market price around a long-run moving equilibrium price by means of a buffer stock, but including export quotas as they were actually imposed by the ITA itself. The study used a number of simulations to determine what buffer stock would have been needed to keep real annual market prices within each of three bands around the long-term trend. The bands were ± 5, 10 and 15 per cent. Narrower or wider limits were ruled out, the former because it would have been too expensive to defend them, the latter because the degree of stabilization was judged to be insufficient. It was found from the simulation that to have kept price fluctuations over the period 1956 to 1974 within the ± 15 per cent range, a buffer stock reaching a maximum of at least

100,000 tonnes would have been required. It was noted, however, that within the more stable period 1966 to 1974, 'depending on the price band-width, stocks of the order of 40,000 (± 15 per cent) to 70,000 (± 5 per cent) tonnes would have had to be maintained'.

A more recent buffer stock simulation study has been carried out by the World Bank, using two assumed price bands of ± 10 and 15 per cent, and the actual price bands chosen by the ITA. The Bank study concluded that over the period 1961–75 the maximum buffer stock required for the ITA range would have been 51,000 metric tons, but for the ± 10 and 15 per cent bands (deviations from the trend) the respective maximum would have been 140,000 metric tons and 103,000 metric tons, respectively. As the study points out, there is no doubt that the results concerning the size of the buffer stock which would be necessary to maintain the integrity of the bank, depend heavily on the price-setting rule chosen by the stabilization agency.[33]

Over the periods covered by these two studies it seems that the Tin Council's resources, certainly the basic producer contributions, were totally inadequate to defend the ceiling. Smith and Schink are quite categorical in their claim that the dominant influence in the market was the GSA, not the Council.

> Producers realised that a push for high floor prices, after the manner of a cartel, would probably have brought the GSA into the market. Consumers, on the other hand, had reason to believe that they could rely on GSA sales to mitigate the largest penetrations of the ceiling. Hence, they did not press as much as they would otherwise have done for larger, more effective buffer stocks . . . These two items of conflict, the height of the floor price, and the size of the buffer stock (and its financing) were thus greatly mitigated by the GSA stockpile.[34]

It is worth emphasizing, in judging these studies, that the need for GSA sales in the early sixties was greatly influenced by powerful non-commercial forces with which a commerically-determined buffer stock could not be expected to cope. The slow recovery of production in the first half of the sixties was due to a large extent to political troubles in Indonesia and Zaire, and to a mixture of political and economic difficulties besetting the run-down nationalized sector of the Bolivian industry. A large amount of tin was inevitably lost in these countries, which had collectively accounted for 44 per cent of world production in 1956. It would be unrealistic to expect an ordinary buffer stock to cope with prolonged difficulties of this kind affecting nearly half the world industry. It is more the kind of situation for which government-financed national strategic stocks might be held as an insurance.

It has been argued that the Council could have made more use of the buffer stock, since the experience of 1957–8 had shown that it was not limited to the money from the basic producer contributions. Gilbert maintained in 1976 that the Council's use of export quotas rather than the buffer stock was only partly due to lack of money. 'The facility for buffer stock borrowing against the security of tin warrants was under-utilized and the further option of bank borrowing was used only modestly and as a last resort measure, even during periods of severe export control'.[35] Criticizing the use of severe export control in the late fifties, a World Bank study, admittedly with the benefit of hindsight, has suggested that if tin foregone by the export cuts had been taken up by the buffer stock, it would have repaid its carrying costs if sold during the years of growing commercial demand in the sixties.[36]

There is no doubt, however, that producers regarded export control as an essential element in the working of the agreement. They must have judged that they would be better off by keeping up the price in a weak market through export cuts, even if they had to sacrifice output for a time, than by financing larger stocks which might have a depressing effect on the market price. In principle, there is nothing wrong with curbing sales of a primary commodity on a weak market. Manufacturers naturally cut their output if sales fall off, although they may have to accept temporarily a rise in stocks. Unfortunately, export controls have raised serious problems for the Tin Council, namely, the timing of their introduction and removal, the extent of the cuts, their incidence in various parts of the industry, and the possible repercussions on future production potential.

In his sympathetic but not uncritical study of the history of the agreement, Fox has referred to mistakes in operating export controls. Looking at the first prolonged period of control, Fox maintains that it was kept on far too long when commercial demand was clearly growing. The reason apparently was to get rid of the tin which had been bought with the help of the special fund, and then to reduce the amount remaining in the buffer stock. In his view, the Council subordinated the interests of production to those of the buffer stock, which meant that employment suffered heavily in the member countries where the industry was privately run, and a large number of mines had to close down. It also meant, in Fox's opinion, that 'the long-term consequences of the retention of export control for too long, and the absence of almost any buffer stock tin from June, 1961, left the market wide open under the second agreement to the US surplus strategic stockpile'.[37] Fox is also critical of the timing and duration of export control in 1968-9, when there was no pressure on the market from US stockpile disposals or sales by the Soviet Union, 'This illustrates, in the later stages of exploding prices, the consequences of miscalculating the right time to get out of export control'.[38] Similarly, it was an official American view that the export control period of 1975-6 was continued too long and had the effect of accentuating the sharp rise in price which followed.[39]

It is naturally difficult to make accurate market forecasts in stabilization operations. Hence it would not be surprising if some mistakes were made in judging the amount of tin that could be kept off the market by export control without adverse repercussions. The problems of manipulating a buffer stock and export control are complex when stock disposals are involved.[40] Several criteria have to be weighed in the balance: maximizing export values, protecting the interests of importers by preventing excessive price increases and those of producers by minimizing long-term substitution effects, and restoring the liquidity of the buffer stock agency. Conflicts have arisen between consumers and producers over these criteria, with consumers reluctant to agree to the renewal of control, and producers, although by no means always unanimous, less willing to take the risk of freeing the market.

How far export control will be required in the future is uncertain, but producers will not give up the option of control. Much depends on the use that might be made of borrowing rights which would swell the financial resources available to the buffer stock. Although borrowing rights have been substantial, producers might well be unwilling to see another large stock hanging over the market.

It is uncertain whether the agreement had any effect on the average price level, but it may have helped to raise it by the short-term and long-term restrictive effects of export control, although this was not necessarily beneficial to the prospects of the industry as a whole. The agreement does not seem to have done anything to stimulate consumption, in which respect tin has clearly lagged behind other metals.

To sum up, what has the agreement achieved over its almost quarter of a century history? Would the tin market have been much different if there had been no more than the international study group such as existed in the earlier postwar years? In the last resort it can be only a matter of judgement. On the positive side must surely be stressed the guaranteed floor price, even if the degree of stabilization was not vastly better than that of other commodities, and due, in part at least, to an outside agency.

There are other considerations, more important now than they were in the fifties, long before there was any talk of a 'new international economic order'. Developing countries resent their lack of influence on international commodity markets where they sell most of their output. Market forces are not left unqualified in any industrial country, as the experience of the European Community and the US with agriculture proves. The international tin agreement has seemed to give developing countries some say in controlling market forces, even if it has fallen well short of the ideal. But there has certainly been restlessness on the producing side, as shown by the reluctance of Malaysia and Bolivia to ratify the third and fifth agreements, respectively, and the difficulties in negotiating a sixth agreement.

The UNCTAD plan for a Common Fund under the Integrated Commodities Programme has included tin as one of the 'core' commodities for which there could be buffer stock operations. When the Common Fund comes into being, the ITC would no doubt be associated with it in some way, since member governments are likely to be also members of the Fund. The draft sixth tin agreement includes under Article 25 the statement: 'When the Common Fund becomes operational, the Council shall negotiate with the Fund for mutually acceptable terms and modalities for an association agreement with the Fund, in order to take full advantage of the facilities of the Fund'. That there might be problems was suggested by the conversion of UNCTAD to the view during the protracted negotiations for the creation of the Common Fund that attempts to encroach on the autonomy of the cocoa, coffee and tin agreements would spell the breakdown of alliances with these commodity organizations.

The profitability of the tin buffer stock and the maintenance of the floor price have been encouraging to UNCTAD, but there must be acute difficulties in reaching workable compromises among the conflicting interests involved in the other 'core' commodities. Even the tin agreement, in spite of many years' experience in making compromises, has been under great strain at times. The problem facing other groups of producing-consuming countries can hardly be underestimated. There could be more promising results, however, from the use of the 'Second Window' under the Fund, with commitments from consuming countries to help in the financing of programmes to improve the efficiency of extractive industries and possibly the marketing of primary commodities. Bolivia, for example, would benefit, although the total sums available to the Fund appear very modest, given the needs of many producing countries over the range of 'core' commodities.

The strains to which the tin agreement has been subject became particularly acute during negotiations for a sixth agreement in 1980–1, and compelled as a stopgap an extension of the fifth agreement, pending the closing of the gap between the sharply diverging interests. The main points of dispute were the size and financing of the buffer stock, the role of export controls and the effects on prices of stockpile disposals.

At the start of negotiations in 1980, the US unequivocally rejected the inclusion of export control in a new agreement and advocated a purely buffer stock arrangement with a capacity of up to 70,000 tonnes, equivalent to about 35 per cent of non-communist world production at the 1980 rate. The producing countries wanted a limit of 30,000 tonnes and insisted that the question of export control was non-negotiable. Bolivia took the extreme producer view, pressing for a still smaller maximum and a low point at which export control could be introduced. The EEC proposed, as a compromise, a normal buffer stock of 35,000 tonnes and a contingency stock of two-thirds of normal stock, financed by loans on the security of stock already held and with government guarantees if necessary, and finally, export control only under strict conditions. Subsequently, the US proposed a somewhat smaller buffer stock, but with export control only under restrictive conditions, including a high minimum stock at which control might be considered. An important gap remained between the US position and even a modified package deal proposed by the chairman of the Council, leaving deadlock after months of negotiations in 1980–1.

The conflict between the US and the producers over a sixth agreement was simply a continuation of previous difficulties. It was also an indication of the obstacles facing the negotiators of even long-established commodity agreements. In this case some producers at least may well have regretted the adhesion of the US to the fifth agreement. The US, in fact, seems never to have overcome its previous ambivalent attitude to the tin agreement, and there remained strong business opposition to membership, especially the steel industry.

In late 1981 it was still uncertain whether the US, and also Bolivia, would eventually ratify the generally agreed compromise sixth agreement due to come into effect in July 1982. Unlike previous agreements the sixth could come into effect with 65 per cent support from producers and consumers; previously 80 per cent of the votes of producers and consumers were needed. The new agreement included a 30,000 tonne buffer stock financed jointly by producers and consumers, and another 20,000 tonnes to be funded, if necessary, by bank loans raised against buffer stock warrants and, if necessary, with government guarantees. Export control could come into effect only if there were two-thirds support from both producers and consumers, and not before buffer stock holdings reached 35,000 tonnes, considerably above the level of previous agreements. Provision was also made for a more rapid moderation of export control when the market price situation improved.

Cartel Prospects

Policy disagreements over the working of the agreement in the seventies revived the idea of a cartel in which the producing countries would 'go it alone'. There was a hint of a new approach in discussions between Bolivia and Malaysia which led to the so-called La Paz Declaration of February 1977.[41] It was followed by the creation of a producers' secretariat to co-ordinate policy, particularly for the negotiation of the sixth agreement. Two points of the Declaration are of particular interest; firstly, high prices will not turn away consumers from tin, either in the long run or the short run, simply because developing countries can be counted on to increase their consumption; and secondly, producing countries must be guaranteed a price high enough to enable them to diversify their economies into other areas of production.

The first point suggests that some producers, at least, believe that there is an element of bluff in the warnings that technological progress is a serious threat to the total demand for tin. Whatever happens on the technological front can be offset by the growth of demand from the sheer mass of potential consumers in the developing countries. Such a view would imply a heavy reliance on the small number of newly industrializing countries out of the much larger number of developing countries, whose tin consumption cannot make much difference to world demand over the next 20 years. This point is elaborated in a later section.

The second point implies that the cost of production should include a large taxable margin. For a large part of the industry, at least, such a margin already exists, in so far as there is a wide spread of costs, which allows governments to tax the economic rent accruing to intramarginal producers. The closer the price is to the costs of higher-cost producers, the larger the taxable potential for the governments of low-cost producing countries. The Bolivian view, apparently, is that the floor price should be high enough to cover COMIBOL's costs, including heavy fiscal charges. From the point of view of producers as a whole, such a policy carries the risk of an excessive stimulation to production, unless other countries reckoned that they would be better off with a relatively high price and lower volume, and could check any response by private producers to the incentive of a higher price.

If there were to be a cartel, its success would depend on a number of conditions: a high proportion of output from participating countries, with outsiders incapable of significantly increasing their output, a low elasticity of demand, and loyalty by participants in carrying out cartel decisions, in other words, no chiselling. Prewar experience showed that tin could satisfy these criteria, although there were problems over the allocation of quotas, and outsiders caused some difficulties. At the end of the seventies, the four leading producers accounted for about 80 per cent of world production, excluding China and the Soviet Union, neither of which seemed likely to be a threat. The new producer, Brazil, showed no signs of rapid expansion. Its mines were deep in the heart of the continent, and in any case domestic consumption was expanding. Even if the two African producers staged a big recovery in the long run, they might be expected to align themselves with other developing countries on cartel policy.

The main checks to a cartel would come from the surplus US stockpile and the reaction of leading industrial firms to what they believed excessively high prices. It is uncertain by how much the average price could be raised without adversely affecting consumption, but there seems to be evidence that the average price in the past has done so. If a cartel were to aim at a higher average price, its objective should ideally be to stay just clear of the 'limit price'. By limit pricing is meant pricing which just makes substitution unattractive, taking into account the capital costs of switching away from tin. Whether a cartel could tune its policy sufficiently to keep below the limit price is debatable, given the difficulty of determining the limit and the pressures to maximize economic rent. A World Bank report came to the conclusion that the 'exaction of monopolistic rents must remain a highly uncertain proposition',[42] but presumably the OPEC would have to be excluded from this pessimistic judgement.

Although the prewar cartel seems to have been successful in maintaining a relatively high price for tin in much of the thirties, it cannot be certain that a similar success could be achieved in the eighties. Apart from the stockpile, the threat of substitution is greater now as a result of the faster rate of technological progress. Looking at it impartially, however, consumers could not reasonably regard as exploitation a price policy by a cartel which simply aimed at a price adjusted for inflation in the industrial world, that is, indexation. If real wages are more or less maintained in the industrial world there is at least a moral argument, as J.D.A. Cuddy points out, for some form of indexation in the primary producing sector.[43] Part of the problem, however, lies in the reluctance of consuming countries to accept the starting price which producers would want as the base for indexation. This was the source of confusion in relations between the OPEC and the industrial oil-importing countries. As one observer has noted, indexation has been used

> as a legitimizing formula by the oil producers to make up through nominal price increases, for the considerable erosion of prices imposed in 1973-74 . . . the bitter resistance of major developed countries to this demand confuses the continued debate on the cartel price as such, and the grounds for adjusting price which would be reasonable if the price itself were an internationally agreed one.[44]

Arguments over prices and costs between tin producing and consuming countries also arise because there is a similar confusion. In addition, of course, consuming countries do not believe that the price mechanism is allowed to work effectively on supply in some producing countries. A cartel would allow producers to make decisions without first running the gauntlet of consumer opposition. That they were willing to ratify five successive agreements suggests that a cartel was still regarded as a second-best alternative.

Notes

1. The other commodities were wheat, sugar, coffee, olive oil and cocoa. Up to the end of 1979 the cocoa agreement was still not in operation. Rubber was added to the list in October 1979, but the success of this latest international agreement remains to be seen.
2. K.E. Knorr, *Tin under Control*, p. 79.
3. See W. Fox, *Tin: the Working of a Commodity Agreement*, esp. Ch. Ten.
4. Fox, op. cit., p. 248.
5. A discussion of the negotiations for the sixth agreement will be found in various issues of *Tin International* throughout 1980 and 1981. Other references are given in the ITC publication, *Notes on Tin*.

6. Fox, op. cit., p. 256.
7. Fox, op. cit., p. 267.
8. Fox, op. cit., p. 273.
9. C.P. Hallwood, 'The Profitability of the buffer stocks operated under the International Tin Agreement', *Resources Policy*, December 1979, p. 277.
10. Fox, op. cit., p. 277.
11. ITC, *Notes on Tin*, May 1979, quoting a statement by the US counsellor for economic and social affairs at the US embassy for Kuala Lumpur, reported in *Business Times*, May 1979. The same US authority went on to argue that 'various land and taxation policies adopted by the tin producing countries, although meant to attain other national objectives, have unintentionally served to restrict the supply of the commodity'.
12. Statement by the head of the Bolivian delegation to the Tin Council.
13. According to A. La Spada, in a paper presented to the fifth World Tin Conference in Kuala Lumpur in October 1981, the results of the Economic and Price Review Panel were disappointing. Its task was 'to assess short- and medium-term world tin supply/demand position and movements in the price of tin and to analyse other relevant economic indicators in order to reach "findings and conclusions" on the appropriateness of the ITA price range. Consensus on "finding and conclusions" within the Panel was reached on only one occasion and the revision of the price range soon reverted to the "horse trading" technique'.
14. Christopher Gilbert, *The Post-War Tin Agreements and their implications for Copper*, Commodities Research Unit, London, 1976. See also the summary in *Tin International*, July 1976, p. 233.
15. *Tin International*, July 1976, quoting Gilbert, op. cit.
16. J.D.A. Cuddy, *International Price Indexation*, Saxon House, D.C. Heath and Lexington Books, Mass., p. 59.
17. The building up and partial running down of the American stockpile are discussed at length by W. Fox, op. cit., Ch. Eleven and Fifteen.
18. Knorr, op. cit., p. 180.
19. Ibid., p. 180.
20. Fox, op. cit., p. 242.
21. Fox, op. cit., p. 348.
22. Gordon W. Smith, 'US Commodity Policy and the Tin Agreement', in David B.H. Denoon, *The New International Economic Order: a US Response*, Macmillan, London, 1980, p. 194. Smith argues that the erratic behaviour of US stockpile policy increased the uncertainty facing tin producers and stockists. 'Recent shortfalls in tin production [i.e. up to 1979] may owe a good deal to the natural but mistaken belief that the G.S.A. would continue to sell surplus stock during periods of high prices'. See also on national buffer stocks, Helen Hughes, *World Development*, 1975, Vol. 3, p. 823. 'National buffer stocks, generally maintained for defence purposes, have tended to exacerbate price fluctuations', a point which might well be borne in mind by governments contemplating stockpiles of essential raw materials in the 1980s.
23. In May 1980 the strategic objective was fixed at 42,000 long tons.
24. W. Fox, 'Some Thoughts on US Stockpile Disposals', *Tin International*, August 1973, p. 275.
25. The US has made a donation of 1,500 tonnes from its stockpile to the buffer stock, as part of a voluntary contribution of up to 5,000 tonnes originally pledged by the US, *Tin International*, November 1980.
26. Gordon W. Smith, op. cit., p. 193.
27. Kerstin Backman, 'The International Tin Agreements', *Journal of World Trade Law*, Vol. Nine, 1975.
28. Backman, op. cit., p. 517.

29. *Tin International*, February 1977.
30. *Tin International*, July 1976.
31. Fox, op. cit., p. 395.
32. G.W. Smith and G.R. Schink, 'The International Tin Agreement: a Reassessment', *Economic Journal*, December 1976. These severe critics go so far as to say 'the longevity of the ITA may owe a good deal to its very ineffectiveness', p. 721.
33. World Bank Staff Commodity Paper No. 1, *The World Tin Economy: an Econometric Analysis*, Commodities and Export Projections Division, World Bank, Washington, D.C., June 1978, pp. 33–5. See also the discussion by Smith, 'US Commodity Policy and the Tin Agreement', in Denoon, op. cit.
34. Smith and Schink, op. cit., p. 721.
35. Gilbert, *Tin International*, July 1976. C.P. Brown takes a similar view: 'although the buffer stock funds have conventionally been cited as inadequate for the purpose of supporting prices within the price range, they have been used fully only in the period 1957–58 if account is taken of the manager's additional leeway to borrow in financial markets on the security of warehouse warrants for tin as a last resource measure'. *The Political and Social Economy of Commodity Control*, Macmillan, London, 1980, p. 19.
36. World Bank and IMF, *The Problem of Stabilization of the Prices of Primary Products*, Washington, D.C., 1969, p. 95.
37. Fox, op. cit., p. 280.
38. Fox, op. cit., p. 280.
39. E. Allen Wendt, Director of the US Office of International Commodities, State Department, quoted in *Tin International*, December 1978.
40. A problem arises over the behaviour of privately held stocks. It is likely that some substitution can occur between such stocks and the buffer stock. Private users and stockists may shift some of the cost of holding stocks on to the buffer stock, which must, therefore, affect calculations of the optimum size of the buffer stock. How much substitution goes on, can only be guessed by the buffer stock manager. It is, of course, not costless to the private user, since tin acquired for the buffer stock only becomes available again, at least with certainty, at the ceiling price. Gordon W. Smith believes that substitution may have been very significant at times: 'one apparent sign has been the relative ineffectiveness of massive GSA-ITA sales in reducing prices to more reasonable levels in 1964–65 and 1973–74, and the rapidity with which the Tin Council's 20,000 tonne buffer stock was exhausted in 1976'. US Commodity Policy and the Tin Agreement, in Denoon, op. cit. See also the discussion by Paul Holland, *The Stabilization of International Commodity Markets*, Jai Press, Greenwich, Conn., 1979, Ch. Six.
41. The so-called La Paz Declaration was discussed in *Tin International,* March 1977.
42. World Bank Staff Working Paper No. 354, *Development Problems of Mineral Exporting Countries*, Washington, D.C., 1979, p. 95.
43. J.D.A. Cuddy, *International Price Indexation*, Saxon House, D.C. Heath and Lexington Books, Mass., 1976, p. 59.
44. Wolfgang Hager, 'Administrated Commodity Markets: the Search for Stability', in Karl P. Sauvant and Hajo Hasenpflug, *The New International Economic Order: Confrontation or Cooperation between North and South*, Wilton House Publications, London, 1977, p. 162.

9 RESERVES AND FUTURE SUPPLY POTENTIAL*

Problems of Measurement

There has been much pessimism since the 1973 oil crisis about the long-run prospects for maintaining adequate supplies of essential materials for a resource-hungry world. Arguments that the world is likely to exhaust some minerals within an inconveniently short time have received an attentive hearing from the public. In spite of attempts by apparently reliable authorities to refute these views, the climate of opinion is inclined to favour pessimistic forecasts.

At various times over the last 30 years, world tin consumers have needed reassurance about the long-run reliability of tin supplies. Their doubts have arisen from a number of factors, geological, political and economic. It may be felt that tin miners will simply run out of workable reserves, so that it will be impossible to maintain a large enough output in the future. A corollary of this situation would be unreasonably high prices. Sometimes the fact that most tin deposits occur in developing countries in politically unstable areas creates doubts about the chances of being able to look for and exploit new reserves to replace those that are running down.

One sure way of reaching a gloomy conclusion about mineral reserves is to take the so-called 'proved' reserves reported by mining companies and express their sum total as so many years' consumption at a particular rate. This is an entirely mistaken approach to the subject. Mining companies do not attempt to 'prove' their reserves beyond a very limited period. They are in business to make money. 'Proving' additional reserves costs money, which cannot bring any return until

---

* The most up-to-date survey of tin reserves is to be found in the Tin Council report, *Tin Production and Investment*, already referred to in earlier chapters. This chapter relies heavily on the section of the report which deals with reserves. An earlier Tin Council report, also referred to, dealt in some detail with reserves as seen in the early sixties. This chapter also draws on material summarized in various issues of *Tin International*, on the report by C.L. Sainsbury, *Tin Resources of the World*, published by the US Geological Survey, US Department of the Interior, 1969, and also on a general review of published information on reserves by W. Fox, *The Reserves and Availability of Tin and the World Consumer*, a paper delivered to the World Conference on Tin Consumption, London, 1972.

some remote future date. From their point of view, there is no reason to define the extent of their workable reserves beyond some commercially reasonable date. As one mining spokesman put it, 'mining companies are not in business to provide intellectual comfort to analysts of resource availability'.[1]

Many small mines in southeast Asia have operated for over 20 years with never more than five years' proven reserves. Some large mines have been producing tin for many generations, for example, the Catavi mining complex in Bolivia. As we pointed out earlier, leading mining areas have been producing tin for centuries, and one important producer has returned in recent years to the peak outputs of over a century ago.

Whatever the problems and the uncertainties surrounding the question of mineral reserves, it is desirable to build up a picture of the reserves position for individual countries and for the world as whole, or at least for the non-communist world. There is an obligation on the mining industry to keep governments informed of the supply potential, which may be important for both national and regional employment and income prospects. Consumers need to be convinced that they should have no doubts about getting all the tin they want at reasonable prices. 'A consumer kept ignorant of the availability position of tin may be already one step in the way to becoming a future consumer of a readily available substitute material'.[2] Moreover, if consumers and governments are misled by a lack of information on reserves, they may devote resources to the development of alternatives long before it is economically justifiable.

The first problem in examining the state of world tin reserves is the lack of consistency in classifying the deposits. Since different methods of classification have been used, it is difficult to arrive at acceptable aggregate estimates. A broad distinction can be made between 'reserves' and 'resources' along the lines adopted by the US Bureau of Mines and the US Geological Survey.[3] The distinction made by these organizations rests on what is known about the geology of the resources and the economic feasibility of exploiting it. Mineral resources are defined as 'all materials surmised to exist having present or future value'.[4] Only part of these resources are put under the heading of 'reserves', namely the part which has actually been identified. The greater part is assumed to exist according to geologic evidence, the accumulation of which over time may be expected to increase 'total resources'. The amount of reserves in that total may also be expected to change, not necessarily always to increase, because it depends on both economic conditions and changing technology.

Under the heading of 'reserves' a number of commonly-used terms of classification appear in various estimates of tin reserves. 'Measured' and 'proven' are often applied synonymously to reserves. 'Probable' and 'possible' may be related very loosely to 'indicated' and 'inferred', but are far from being synonymous. Sometimes the term 'demonstrated' is used to mean the sum of 'measured' and 'indicated' reserves.

Such differences inevitably mean that estimates of reserves are even more doubtful than geological and economic uncertainties make them, and these uncertainties are undoubtedly formidable, as the recent Tin Council study has emphasized. 'Any figure or estimation which may be put forward for world tin resources must be suspect, therefore, on grounds of both definition and of knowledge, and any time scale attempted as to their possible exhaustion must also be subject to the severest provisos and limitations as to accuracy'.[5] From the economic point of view, there is at any time a certain amount of tin available for extraction at the current cost-price ratio. If the ratio changes, the reserves position also changes, although we are unlikely to be able to measure it accurately. The more favourable the cost-price ratio, the more tin becomes potentially available. Lower-grade deposits, less accessible deposits, move into the commercially exploitable range. Lower-grade deposits are more abundant than high-grade deposits, simply because of the way in which geological processes have developed in the past. If this had not been so, the output of tin, and of other minerals, could not have expanded as it has done over the last hundred years. But expansion has come through the advance of technology, which may be said to have 'converted rock into ore', as the previous review of ore grades has shown. In order to prevent the real cost of tin from rising, it will be necessary to maintain technological progress in dealing with low-grade deposits, although it may still be possible to find deposits at least as good as, if not better than, the marginal deposits currently being exploited. The Tin Council report takes a rather gloomy view of the long-term prospects for costs: 'future discoveries and tin strikes are generally likely to be either lower-grade deposits and/or geographically less convenient for extraction'. It goes on to suggest, 'whilst it would be foolish to predict that the world will run out of tin in the foreseeable future, it should be anticipated that, excluding the possibility of profitable new discoveries, e.g., offshore, mining and recovery costs are likely to escalate to a greater extent than could be expected from the effects of inflation alone'.[6]

Estimates of Reserves

In the last 30 years there have been numerous estimates of the world's tin reserves. The US Paley Commission produced in 1952 an estimate of 5 million tons for non-communist reserves.[7] The Commission's inquiry was the response to a wave of anxiety in the US about the long-term prospects for meeting America's growing industrial material needs. In 1965 the Tin Council published the results of a general study of the world tin market, including a detailed section on the industry's reserves.[8] This inquiry came at a time when current production was running well below consumption, and consumers needed reassurance about both the short-run and the long-run prospects.

The ITC report excluded communist countries, as well as Burma, a small but growing producer in the thirties, and Brazil, the latest and growing producer in the sixties. Offshore deposits were also left out for lack of information. The method adopted in the report was to ask governments of producing countries to supply estimates of reserves in 1960, 1965 and 1970 on the basis of three prices. The middle price was approximately the market price in 1961–2, the lower and higher prices were about 20 per cent below and above it. The Council favoured the classification of reserves into measured (analogous to proved), indicated (analogous to probable) and inferred (analogous to possible) reserves. It also tried to obtain information on whether the bulk of reserves was in alluvial and eluvial deposits or in lode mineralized rocks and whether the industry would depend on low-grade reserves in the future.

It was an ambitious attempt to build up as good a picture of the reserves position as possible, subject to all the qualifications normally associated with the study of reserves. Not all the replies took account of the possible effects of price on reserves, but those that did showed sharp changes in estimates at the lower and higher prices. Since different countries opted for different methods of assessment, only a rough aggregation could be made. At the upper price the report gave a figure of 4 million long tons in 1965 and 3.5 million tons in 1970. The general conclusion was that reserves should be adequate without a big price increase throughout the seventies and eighties, in short for over a quarter of a century, and it was stressed that the picture was by no means complete in view of the sources excluded from the inquiry. Moreover, 'given the important technological forces at work in the tin-using industries, it was felt that there was no reason to fear that consumers would have to adjust themselves hurriedly to a shortage'.[9] For those who felt that the Malaysian estimate of less than one million tons was ominously low, coming from by far the largest producer, the report pointed out that the prewar Fermor estimate for Malaysia had also been one million long tons, a quantity which had already been mined in 1960.

The next major review of the world reserves position was made in 1969 by C.L. Sainsbury, who extended the coverage of the 1965 report and concluded that world reserves should be adequate to meet demand at the current rate of 180,000 long tons for at least 31 years.[10] Making a distinction between reserves and resources, Sainsbury went on to suggest that the latter were capable of sustaining a consumption rate of 200,000 long tons for another 56 years, beyond which there would be new sources of tin.

Since 1969, there have been other ventures into this part of the forecasting business. The US carried out its first wide-ranging reviews of mineral resources since the Paley Commission. In 1973 the US Department of the Interior and the US Bureau of Mines took a remarkably optimistic view of future tin potential.[11]

Their estimate brought within its scope much more than was usually included. Measured, indicated and inferred reserves were put at 10 million long tons. In addition to these categories, values were put on conditional resources (paramarginal and sub-marginal) and on undiscovered resources (hypothetical and speculative). The former were put at 10,250,000 long tons, the latter at nearly 17 million tons. On the heroic assumption that they could be added up, the report stated that the total of 37 million tons would be 'sufficient to sustain a world consumption of 200,000 tons a year for about 185 years'.[12]

The report emphasized that future tin supplies were most likely to come from areas traditionally associated with tin mining. If new important deposits were to be found elsewhere they could only be in largely uninhabited areas where no significant exploration had occurred, like the new Brazilian tin mining area which had been found in a remote part of South America. It was pointed out that there was no merit on speculating on the finding of major tin deposits in entirely new geologic environments, since tin displays such 'a well-marked geologic association'. It went on: 'each discovery of new tin deposits merely reinforces that fact the deposits containing economic amounts of tin are invariably similar to those deposits which have been mined for thousands of years'.[13] This was a source of optimism in the report. 'To date, no major tin-bearing area of the world can be said to be totally exhausted, as proved by the re-establishment of a significant production in Cornwall [it might have added – and Australia] . . .

since the beginning of historical times, tin has been available except when major wars or other political events have temporarily disrupted its usual flow'.[14] No doubt the industrial importing countries would be happier if they were less dependent on politically unstable sources of supply. Unfortunately, this is a fact of economic life in the post-colonial age. The distribution of reserves, therefore, becomes important. Only two developed countries, Australia and the UK, are confidently expected to have substantial tin reserves, and not on the scale to do more than meet a small part of non-communist consumption. The recent Tin Council study quoted no figures for the UK other than those originally available for the 1965 report, an estimate from the industry of 37,000 long tons in 1970, most of which would be gone by 1980. Since the industry is still producing tin on an even larger scale than in 1964, it is evident that the estimate was relevant only at the time. The US report showed the same optimism about Cornwall that it showed for other long-established mining areas. The well-known authority on the geology of Cornwall, K.F.G. Hosking, also thought that the area was by no means exhausted.[15]

Of the other parts of Europe where tin has been mined, the US report suggested that Spain and Portugal might have 'good possibilities because of the favourable geology, the known deposits, and the archaic state of mining and exploration',[16] but in its view, no other part of Europe, not even the Erzebirge, once an important mining area and extensively explored during the last war, offered any serious prospects.

Australia is a much more promising long-term source. Since the first Tin Council report, estimates of Australia's reserves have greatly increased. The estimate for 1960 was 47,000 long tons of measured, indicated and inferred reserves; a 1973 estimate by W.G.B. Phillips came to 384,000 tonnes, and the latest supplied to the Tin Council by the Australian Bureau of Minerals Resources has risen to 920,000 tonnes, about one-third of which is put in the categories of measured, indicated and inferred reserves.[17] The important Renison deposit, known since 1890 but not mined because of its complexity until 1965, was first reported in 1964 to contain reserves of 5,900 long tons 'fully or partly developed'; a 1978 estimate of proved and probable reserves was 155,000 tonnes, nearly 30 years' output at the current rate.[18]

That Australian reserves have increased so much in the last 15 years, and more than offset the amount extracted during that period, reflects a more rigorous compilation of previously known deposits and an intensified degree of exploration by companies with higher rates of production than in the sixties. The fact that such a stable area has good long-term prospects is obviously encouraging to importing countries. Even so, it is apparent that the scale of Australian reserves could not support anything like the lack of output which the large southeast Asian producers have maintained for many years.

As by far the largest producer, Malaysia is particularly important in any assessment of reserves. There is, however, relatively little information on the overall potential of the Malaysian deposits, none at all on the prospects for lode mining which may be possible in the distant future. The shortage of published information on Malaysia's alluvial deposits, which have been the source of practically all the country's tin, is due, not to inadequate exploration and prospecting, but rather to the difficulty of collating the findings of many different agencies. The fragmentation of the industry in the past, with several hundred companies of varying sizes carrying on their own prospecting to a greater or lesser degree, discouraged official attempts at aggregation of reserves. The Malaysian reply to the Tin Council's questionnaire in the early sixties gave a figure of 900,000 long tons. That it was a very conservative figure was proved subsequently by a total output of 910,000 tonnes by 1980. In response to the 1979 Tin Council inquiry, the Malaysia delegation reported that at the current rate of exploitation some 34 years of life remained, which implied reserves of about 2 million tonnes. The estimate, however, does not appear to cover offshore areas. There has also been a large increase since 1970 in the total acreage under prospecting rights. Although the acreage actually selected for mining has fallen, it is possible that an increase in prospecting could reveal more deposits. That major discoveries could be made was shown by the discovery of the Kuala Langat deposit at a greater depth than normal in Selangor State in 1973. The 5,000 acres likely to be mined initially, out of what is believed by some to be a vast untapped tin reserve, would involve the biggest investment in the tin industry for many years. Unfortunately, the Tin Council report could give no details of grades. It would be useful to know if this new ground is better than currently mined ground. Uncertainty over this and the reserve potential in the largest producing country tends to perpetuate a lack of confidence among major users in the importing countries, and to encourage the search for substitutes.

Second only to Malaysia is Thailand. As a result of the expansion of its industry in recent decades, there are expectations that it could have a great potential. The Tin Council report quotes an estimate of 863,000 tonnes. However, this is well below the figures given in response to the previous inquiry. In contrast, estimates of Indonesian reserves have increased since the early sixties. This is not unexpected, in view of the increase in exploration by the state mining corporation in the last ten to twelve years. The estimate given to the first Tin Council inquiry was 660,000 tonnes. The updated 1977 estimate was 1,640,000 tonnes, of which over one million was classified as measured, indicated and inferred reserves.

An increasing proportion of Indonesian reserves appears to lie offshore, according to the trend of production. The state corporation believes that over 40 per cent of current reserves already are offshore. Presumably this excludes any tin which may lie beyond the depths currently accessible to offshore dredges. Thailand also may have substantial reserves offshore, as the large increase in small-scale suction boat mining suggests. In both countries, however, there is clearly a need for a new technology to enable the mining of deeper offshore deposits, which implies some degree of participation by large foreign mining companies.

Technological progress is also a condition for the development of new tin deposits in Nigeria. The Tin Council report quotes a 1972 estimate of 100,000 tonnes of proved reserves for surface mines. Recent information on both Nigeria and Zaire is scanty. Fox has suggested that Zaire contains 'one of the great tin reserves of the world and it would appear reasonable enough to regard its mining potential as having been little more than scratched'.[19] The figures available to the Tin Council in 1979 were only 45,000 tonnes of proved reserves and about 100,000 tonnes of probable and possible reserves. The US estimate was larger, but not significantly so for measured and indicated reserves. Zaire, in fact, is more of an unknown quantity now than it was 20 years ago. It is evident from the declining output figures that the industry is in great difficulties. Probably very little is being done to find additional reserves. The Tin Council report gives an estimate of US$40 million as the sum required for hard-rock mining in one important tin field. The general impression of the position in the two African countries is of possible substantial reserves, but not much prospect of a major contribution to world output for a long time, and certainly not without the assistance of foreign companies.

The 1979 Tin Council report does not give a comprehensive estimate for Bolivian resources, partly because so little is known of the reserves of the 5,000 small mines. Exploration and prospecting have been handicapped in Bolivia by lack of money for many years. It is likely, however, that the position may be improved in the eighties as a result of the creation of the National Mining Exploration Fund, financed by the government and several foreign sources.

The US report gave an estimate of about one million tons in measured, indicated and inferred reserves, with much more under the heading of resources, as defined in the Bureau of Mines Classification system. The Tin Council quotes as estimate of reserves only for mines managed by COMIBOL, accounting on average for about two-thirds of Bolivian production. The figure of 500,000 tonnes is equivalent to about 25 years' output at the 1978–9 rate. There is no reason to believe, however, that this figure cannot be increased with more prospecting. Much tin still lies in the vast tailings which have accumulated over the last hundred years at the larger mines. Nor is there reason to doubt that the private miners will be able to continue working, as they have in the past, with never more than a small amount of reserves. And finally, the great size of the Bolivian tin belt, compared with the African or even the Malaysian deposits, gives reasonable hopes of future discoveries. Whether the fall in the grade of ore can be halted or reversed is uncertain. This would be important for the future trend of Bolivian costs.

The other important producing country, Brazil, is believed to be only on the threshold of its history as a tin producer, with only a small part of the Rondonian field having been explored. The US report suggested that Brazil could be put on a par with Malaysia, but this was certainly looking far ahead. The Tin Council showed optimism but in a more guarded way.

The Tin Council has not attempted a global estimate of non-communist reserves by adding up all the figures for various countries. It simply concluded that on the basis of the figures, which were the best known to it in 1979, there are 'adequate reserves to meet world demand for at least the next 30 years'.[20] This seemed to the Council a reasonable degree of assurance of supply prospects. But for the longer term it added the judgement of Hosking, that

> we can feel sure that we have a breathing space, perhaps from, say 50 to 150 years, perhaps considerably less than the lower figure, during which our house must be put in order. To do this we must endeavour to locate without undue delay, not only those deposits that we can exploit by known techniques, but also those others that will have to await exploration until it is necessary to tackle

very low grade, or otherwise difficult deposits, and/or until technological break-throughs make them economically viable.[21]

If it can be assumed that the mining industry responds to the challenge of demand, there should be tin resources for an indefinite future. The fact that reserve estimates are repeatedly shown to be too low by subsequent experience of production is one pointer to this conclusion. Even if very low-grade ores must eventually be the main source of tin, one would have to assume that the industry could not find the technology to cope with them if a pessimistic view of the future were to be justified. However, the crucial issue is the trend of costs, whether tin is available only at ever increasing real costs, in which case one would expect substitution to check demand. The pessimistic view is that the future must be one of growing resource stringency with rising costs preventing the growth of the world economy, and probably even reversing it. This implies that ultimately substitution does not work because scarcity is too widespread. Without going further into this general debate on the alleged eventual scarcity of resources, it can be argued that, as one of the most expensive metals with clear signs of rising costs, tin is a pointer to the future. Since not much is known about the grade of future reserves, the pessimist might take the view that they are very likely to be lower than currently mined deposits. Again this is not necessarily true. They might be deeper deposits, which would involve some additional costs, but only if no technological progress were made. However, as a previous section has shown, there are other elements in costs. Energy costs might be higher. Wages might be higher if the producing countries had more developed economies. In this case wage levels would rise as the opportunity cost of labour rises. Whether wage costs rose would depend on the productivity of labour, which brings us back again to technological progress. In this respect the growth of offshore mining in southeast Asian waters is an encouraging development.

Offshore Deposits

Offshore mining is expected to play an increasing role in the industry in the three main southeast Asian countries. So far it has been confined to Indonesia and Thailand, where it has been going on for many years. Including the output of the suction boats in Thai waters, offshore operations accounted for about 15 per cent of non-communist production in 1978. Part comes from small-scale operations, part from very large modern dredges with high labour productivity compared with productivity in much of the industry. Not much information is available about the extent of offshore reserves, where prospecting work, even with the most modern equipment, meets unusual difficulties.

A limiting factor so far in offshore operations is the maximum depth at which even the latest dredges can work. While offshore dredging is highly successful in shallow (less than 50 metres) and reasonably calm waters, it is still not practicable nor cost effective at depths in excess of 50 metres and in rougher waters. Various techniques have been proposed to overcome this limitation on pushing the frontier of offshore dredging out to new recoverable reserves. The current state of thinking on the problems of offshore dredging has been reviewed by Ian A. Knights.[22] Knights points out that 'the basic change that seems to be required, to overcome both the water depth and the surface environmental restrictions, is to structurally isolate the mining device from the surface treatment-transfer vessel . . . the mining device can then be relatively independent of surface vessel marine motion characteristics'.[23] Knights goes on to discuss the relevance of the offshore oil industry for tin mining. 'With the development of commercial seabed vehicles for offshore oilfield operating, the means for implementing this approach now seems closer, without involving the enormous commitments of resources required for more fundamental departures, as, for example, in deep sea manganese nodule recovery'.[24]

There seems to be a good prospect that eventually there will be a successful spillover from the great advances being made in offshore oil technology into offshore tin mining in southeast Asia. One technique under study by a leading British research team is the Drill Stem Miner, a rigid system controlled from the surface and involving motion compensation equipment. This equipment, according to Knights, is believed capable of operating in depths from 30 to 200 metres all the year round in rough water locations from either a new barge or ship-shaped vessel, or from converted existing vessels. It is suggested that such a new technique might make mining beyond the existing limits a commercial proposition by the end of the eighties. Similarly, according to a leading mining authority actively engaged in offshore operations in southeast Asia, new techniques being sought for deeper offshore work might be commercially viable by the late eighties, subject to the availability of substantial mining areas and satisfactory royalty and tax arrangements for private enterprise.[25]

That much research is going on in offshore work is a pointer to future innovations in tin mining, where the basic techniques have not changed since the introduction of dredging on a large scale in the twenties. To what extent the reserves position and the cost of mining with more capital-intensive techniques would be affected is still uncertain.

Notes

1. Quoted in *Tin International*, May 1976, p. 158.
2. W. Fox, 'Tin Reserves and Availability of Tin and the World Consumer', a paper delivered to the Conference on Tin Consumption, London, 1972, published in the *Proceedings*, ITC, London, 1972, pp. 21–45.
3. ITC, *Tin Production and Investment*. The US definitions are discussed in the Appendix to the report, pp. 161–4.
4. ITC, op. cit., p. 11.
5. Ibid.
6. Ibid.
7. William S. Paley, *Resources for the Future*, Materials Policy Commission of the President of the United States, Washington, D.C., 1952.
8. ITC, *Report on the World Tin Position*, 1965, pp. 80–92.
9. ITC, op. cit., p. 160.
10. C.L. Sainsbury, *Tin Resources of the World*, Geological Survey Bulletin 1391, US Geological Survey, Washington, D.C., 1969, p. 10.
11. *Tin International*, June 1973, America's second Paley Report.
12. *Tin International*, op. cit., p. 185.
13. Ibid.
14. Ibid.
15. K.F.G. Hosking, 'The Search for Deposits from which Tin can be profitably recovered now and in the foreseeable Future', a paper to the Fourth World Conference on Tin, Kuala Lumpur, 1974, published in Vol. One of the *Proceedings*, ITC, London, 1974, p. 71.
16. *Tin International*, op. cit., p. 185.
17. ITC, *Tin Production and Investment*, p. 14.
18. At the end of June 1980, Renison's proved and probable ore reserves were put at 15.66 million tonnes at a grade of 1.11% Sn. This represented an increase of 707,000 tonnes over the previous year's total, albeit at a somewhat lower grade. Possible ore reserves at the same date amounted to a further 11.15 million tonnes at a grade of 1.05% Sn. Reported in ITC, *Notes on Tin*, January 1981, quoting the *Mining Journal*, 5 December 1980.
19. W. Fox, op. cit., p. 13.
20. ITC, *Tin Production and Investment*, p. 30.
21. ITC, op. cit., p. 30.
22. Ian A. Knights, 'Deep Water Tin Mining', *Tin International*, August 1980, pp. 316–20.
23. Knights, op. cit.
24. Knights, op. cit.
25. Paper presented to the Fifth World Conference on Tin, Kuala Lumpur, October 1981, by Billiton International Metals BV, Netherlands. This paper dealt with the criteria applicable to investment in tin mining on a large scale and of a capital-intensive nature, with particular reference to offshore mining.

10 FORECASTING MARKET TRENDS

Methods of Forecasting

Commodity forecasting became more common in the seventies, partly as a result of the greater uncertainties which were felt in the industrial countries about the future supply of important minerals, partly with the improvements in econometric techniques and the desire of practitioners to make the most of them.[1] Econometric models have been constructed to forecast prices, consumption, production and inventories. These models give forecasters a mathematical framework with which to examine the complex interrelationships constituting a world primary commodity market. Some models may be used to study the working of commodity agreements by simulating a buffer stock operation over a past period. In this way an estimate may be made of the size and monetary-commodity composition of the stock which would be necessary to achieve certain price objectives, for example, keeping the market price within certain specified limits of the long-term trend or the equilibrium price.

Various techniques have been employed for commodity forecasting. Resource base models use geostatistical analyses of mineral deposits, including probabilistic examination of the relevant resource base linked to supply and demand in a system framework. Various interactions are postulated between the resource base, the rate of depletion of the base, exploration, investment, and the rate of production and consumption. Market models use a dynamic micro-econometric system, making various assumptions about the interaction between decision-makers in reaching market equilibrium between production and consumption, the size of inventories, prices, trade flows and other factors affecting the market. There are also qualitative forecasting methods, largely non-mathematical in their approach, which may, like the Delphi method, take the views of market experts, perhaps on both the demand and supply sides, and construct a forecast embodying their judgements.

All these models face serious problems in coping with the complex reality of commodity markets. In forecasting prices, for example, the behaviour of inventories causes difficulties. Both decision-makers in the market place and forecasters are handicapped by the lack of comprehensive and reliable information on inventories. The Tin Council has had practical experience of this problem. Uncertainty over inventories means that the forces governing price expectations may be very unstable at times. For this reason, argues E.C. Hwa, 'unless the variability of price expectations can somehow be captured, forecasting commodity prices (in the short run) will be a very risky venture'.[2] Over the long run there are also serious problems, as Tilton points out in the study of the US demand for tin. Referring to the possible impact of technological change on the future demand for tin, Tilton comments: 'Materials substitution greatly complicates the task of forecasting mineral requirements, and is likely to render any technique, regardless of the category to which it belongs, vulnerable to wide margins of error when forecasting over the longer term, say 10 to 20 years into the future . . . and even in the short run there can be quite sudden changes'.[3] In Tilton's opinion, materials substitution is likely to become even more important, as firms try to evade rising raw material costs, and technological developments make switching between materials easier. A good example of the latter would be dual-purpose can-making equipment, which enables the rapid switching between tinplate and aluminium in response to changes in relative prices.

To some extent, aggregate forecasting of demand would be helped by the fact that materials substitution does not occur across the board and is not uni-directional. Substitution, therefore, may work to both the advantage and disadvantage of tin. Errors may also tend to cancel out in consumption forecasts, but it is unwise to rely on this outcome.

On the supply side, government intervention in the mining industry of developing countries raises increasingly serious problems, altering perhaps the parameters for the supply functions of parts of the world industry. This is a factor which is much more important than it was in the days when the industry was generally in private, and largely Western hands.

Tin Consumption Forecasts

The world tin market has attracted its share of commodity forecasting, going back at least to the overoptimistic demand forecasts of the US Paley Commission in 1952.[4] Some studies have been much more qualitative than others. There is, for example, a great difference between the approach adopted by the Tin Council report in 1965 and the World Bank econometric study of the world tin economy in 1978.[5]

The World Bank team uses a disaggregated supply-demand model with a market clearing equation for prices, involving 23 behavioural equations. Demand equations are estimated for the US, Western Europe, Japan, South Africa, other developed and developing countries as separate groups. The demand equations take into account the prices of complements and substitutes, but it is admitted that, on the supply side, a number of important factors could not be included because of the lack of consistent information. Among these omissions were government policies in developing countries, labour market problems and taxation. Secondary tin had to be excluded from the model because few countries published data on either their production or consumption. Both the US stockpile and the International Tin Agreement, however, were built into the model.

The predictive potential of the World Bank model was tested by a series of ex-post simulations for the 1955-75 period as a whole and for two sub-periods within it. Remarkably good results were obtained for prices, production and consumption. Assuming, therefore, that the model could capture most of the behaviour of the world tin economy, the World Bank team used it to assess the medium-term prospects for tin up to 1985, with three different sets of assumptions for the exogenous variables: firstly, a high rate of inflation (at least as seen at the time), namely 6.5 per cent, a high growth rate of GNP in the industrialized countries, 5.0 per cent, and heavy US stockpile sales; secondly, moderate inflation, 5.5 per cent, a moderate GNP growth rate, 4.0 per cent, moderate stockpile sales; thirdly, 4.5 per cent inflation, a 3 per cent GNP growth rate, and no stockpile sales.

The results of short-term projections for 1978 and 1979 were less satisfactory than the historical simulation results. On a correct assumption of no stockpile sales, the projection for supply in 1979 was 174,000 tonnes. Actual production was 202,400 tonnes, 15 per cent greater than the projection. The difference between projected and actual production in 1978 was even greater, 165,000 tonnes compared with 198,000 tonnes. The projected output for 1981, however, was almost reached in 1979.

The shortfalls in the model for the two years closest to the completion of the study seem to have been due to the underestimation of the response of some producers after 1977 to a sustained high price. The model did not cope with a very sharp increase in output from Thailand, chiefly due to the activities of the suction boat operators, nor with a somewhat faster rate of growth of output in the Indonesian offshore sector, which benefited from the addition of new capacity.

The consumption projections for 1978-9 were much closer to the actual outcome, 181,000 tonnes as against 171,500 tonnes, and 176,000 tonnes as against 172,000 tonnes. After falling to 170,000 tonnes in 1980, the projections from the model rise year by year to 204,000 tonnes in 1985. The general conclusion of the study was moderately comforting to producers.

> On the whole, it would appear from the projections that tin producers can look forward to a period of slow, but steady rise in import demand as well as to prices that will continue to rise in real terms along their long-run trend. However, as a result of the lags that characterise the response of supply and demand to changing market conditions, the current price boom will be followed by a period of much lower real prices, which is likely to begin in the early 1980s and to last until the mid-1980s.[6]

After holding up well in the late seventies and into 1980, the price weakened markedly in the last quarter of 1980 and the first quarter of 1981. As the Bank study had forecast, the real price of tin certainly fell at this early stage in the decade.

The Tin Council report on the prospect for production and consumption is, like the Paley report, only of historical interest in the 1980s. At the time it was prepared, the early sixties, the tin industry was in considerable difficulties, and it was felt that consumers needed some reassurance about the short-term and medium-term supply prospects, as well as about the longer-term reserves position. The report was largely based on replies to a series of questionnaires by a large number of producing and consuming countries. The projections, therefore, depended heavily on the quality of the returns from participating governments, which varied greatly in the degree of optimism or pessimism about production or consumption in their countries in the projection years 1965 and 1970.

The US forecasts of consumption proved to be excessively optimistic, largely due to an overestimation of the demand for tin by tinplate producers. Since the US was by far the largest tin consumer in the early sixties, its estimates had a major effect on the aggregate consumption projections. The report's estimates for 1970 were a non-communist consumption range of 184-200,000 long tons compared with an actual 176,000 metric tons, and a production range of 163-183,000 long tons compared with an actual 184,000 metric tons. Production did better than expected for several reasons: after giving a pessimistic reply to the questionnaire, Malaysia, the largest producer, subsequently achieved a high level of output as a result of the buoyancy of the gravel pump sector of the industry; there had been difficulties in forecasting the future behaviour of the nationalized industries in Bolivia and Indonesia, both being beset by serious problems in the early sixties; and there was an underestimation of the response of Thailand's industry to relatively high prices and freedom to expand.

Two forecasts of US consumption in the year 2000 were made in 1971 and 1972. The US Bureau of Mines gave a range of 71,000 long tons to 98,000 long tons, compared with an average of 57,000 long tons in the years 1968–70. An interim report of the US National Commission on Materials Policy in 1972 came up with a tentative figure of 90,000 long tons.

Since these forecasts were made in the early seventies, the industrial world, and indeed virtually all countries, have suffered a series of shocks which have inevitably cast a shadow over economic growth prospects and considerably affected thinking on the future demand for minerals. The change is shown in several studies of tin consumption. A report by Wilfred Malenbaum reduces the earlier figures of US consumption for the year 2000 to 67,000 tonnes (see Table 10.1). For forecasts of US and other countries' consumption of tin and other important minerals, Malenbaum takes three factors as the most suitable determining variables, namely, intensity-of-use, the growth of GDP and population.[7] By intensity-of-use is meant the physical amount of tin used per billion dollars of GDP in a particular period. According to Malenbaum, the strength of the forces making for reductions in this parameter are now greater than they were in the early seventies. Moreover, it must be accepted that 'technological advances are adopted readily, even at relatively low levels of GDP per person'.[8] Malenbaum points out that 'tin is the only case studied in which a technological breakthrough may well have dominated the world picture for intensity-of-use, in poor as well as in rich lands',[9] but his projected rate of growth for tin consumption seems to imply a judgement that the loss of ground in tinplate will be less important in the future. As Table 10.3 shows, his 1.9 per cent rate of growth is appreciably higher than the historical rate between 1955 and 1976. Other projections give a similar picture (see Table 10.2). Malenbaum concludes, however, that, of the 12 major metals studied, tin would continue with the lowest rate of growth of consumption for the rest of the century, nor does he envisage much change in the developing countries' share of world consumption; the 'poor nations' would increase their share from 13

Table 10.1: Actual and Projected World Tin Consumption by Area or Country (000 tonnes)

	1971-5	1985	2000
Western Europe	68.84	84	111
Japan	32.76	47	62
USA	52.56	59	67
ODL[a]	11.82	15	22
Eastern Europe	17.14	23	32
Latin America	7.32	11	16
USSR	18.40	24	31
Africa	1.78	7	10
Asia	7.62	11	14
China	14.48	20	28
World	232.72	301	393
Rich nations	201.52	252	325
Poor nations	31.20	49	68

Note: a. Other developed lands — Australia, Canada, Israel, New Zealand, South Africa.

Source: Wilfred Malenbaum, *World Demand for Raw Materials in 1985 and 2000*, McGraw Hill, New York, 1978.

Table 10.2: Growth Rates in World Demand for Selected Non-fuel Minerals (annual averages over periods indicated)[a]

	Historical Averaged over recent 15-20 year periods %	Projected Range of projection 15-25 year periods to 1990 or 2000 %
Tin	1.0	1.3-1.9
Nickel	6.5	2.1-5.1
Refined copper	3.9	2.1-4.0
Iron ore	3.6	2.1-3.2
Aluminium	7.3	3.0-6.7

Note: a. Covers non-communist countries.

Source: Raymond F. Mikesell, *New Patterns of World Mineral Development*, British-North American Committee, 1979.

per cent in 1971-75 to only about 16 per cent in 2000'.[10] Non-communist demand in 1985 is put at 234,000 tonnes, compared with the World Bank team's estimate of 204,000 tonnes. The Malenbaum figure for the year 2000 is 302,000 tonnes compared with an average of 172,000 tonnes in 1978-9. If the centrally-planned countries are included, for which estimates are necessarily even more uncertain, world consumption in 1985 is put at 301,000 tonnes and in 2000

Table 10.3: Historical and Projected Rates of Growth in World Demand for Tin (annual averages for periods indicated)

	%	
Historical	1.0	(1955-76)
US Bureau of Mines	1.5	(1973-2000)
World Bank[a]	1.3	(1974-90)
Malenbaum	1.9	(1975-2000)
Chase Econometrics	2.0	(Ex. US 1979-90)

Note: a. Excludes centrally-planned economies.

Sources: Raymond F. Mikesell, *New Patterns of World Mineral Development*, British-North American Committee, 1979; ITC, *Notes on Tin* April 1980, quoting *Metal Bulletin*, 18 April 1930; Wilfred Malenbaum, *World Demand for Raw Materials in 1985 and 2000*, McGraw Hill, New York, 1978.

393,000 tonnes. It is emphasized, however, that 'in the last analysis systematic consideration of these guides and possibilities needed to be supplemented by the judgment of the research worker'.[11]

Another recent study of tin consumption, by Chase Econometrics, also takes a more sober view of future US consumption than the forecasts of the early seventies.[12] Its estimate for 1990 is only 44,000 tonnes, compared with an actual 48,000 tonnes in 1978. Other non-communist consumption is forecast to grow at an average rate of 2 per cent during the eighties, which would help to offset the virtual stagnation or decline of US consumption. Japanese demand in 1990 is projected to match US demand, involving a 50 per cent increase over the 1977-8 level. Chase put net imports by the centrally-planned countries at only 5,000 tonnes annually on average in the eighties, somewhat less than the actual average in 1974-5, and surprisingly low in view of the higher level of net imports since then.

A medium-term forecast in the Tin Council paper by Yahya and Hollands, based on widespread inquiries throughout the world, arrives at a more conservative and more realistic figure than those of the Malenbaum and World Bank studies for 1985.[13] The Tin Council paper's forecast (excluding the USSR, the GDR and the People's Republic of China) is 190,000 tonnes, some 7 per cent less than the lower World Bank projection, but slightly higher than the average level of consumption in the period 1977-9.

Tin Production Forecasts

Both the World Bank and the Chase reports make projections for production in the next five to ten years. The Chase report suggested Malaysian output might reach its peak about 1982, falling to only about 54,000 tonnes in 1990. Indonesia was forecast to continue its slow expansion of about 2 per cent a year on average to reach 36,000 tonnes by 1990, a rather low estimate given the rate of increase in the last three years. These two countries together, according to the Chase forecast, would apparently produce about 2,000 tonnes less in 1990 than in 1979. The World Bank projection for 1985 is a total non-communist output of 230,000 tonnes, about 30,000 tonnes higher than in 1979. A more recent private US report on production prospects for 1985 gives the rather smaller figure of 216,000 tonnes.

The forecasts of consumption by the end of the century imply a substantial increase in production. By past standards of other non-ferrous metals, and indeed of tin itself, an increase of up to 50 per cent in output over a 20–25 year period is not exceptional. The historical record has shown a vigorous response by producers to a buoyant long-term growth in demand. Production of copper, for example, rose by 40 per cent between the mid-fifties and the mid-sixties. Tin production rose by 55 per cent between 1900 and 1913, and again by 45 per cent between 1920 and 1929. These were not periods in which there was spare capacity to facilitate a rapid growth in output, such as happened during the forties. Lead and zinc have also shown substantial increases in output over comparable periods. However, the absolute size of the additional amount required would now be considerably greater, and the question arises, therefore, of the tin mining industry's ability to achieve a large enough expansion in the future, given the changes in structure and organization which have occurred in the last three decades.

Assuming that consumption were to develop as the forecasts imply, what are the prospects for expansion of output? The industry has four ways of producing more tin: to expand existing mining areas and the associated processing plant at the mines (there should be no serious problems over smelting capacity); to open up new deposits, onshore or offshore, using the most modern and efficient equipment and technology; to improve the recovery rates at mines and processing plants; to re-treat old tailings and mine waste dumps, some of which contain very large quantities of tin. Supply could also be supplemented for a time by sales from the US stockpile. It might also be possible eventually, although the prospects are not very encouraging, to recover some secondary tin from used tinplate containers.

Considered in the context of the world mining industry, the total investment required to create additional tin mining capacity is not enormous. Mikesell estimates that the capital cost per metric ton of annual tin producing capacity lies in the range of US$10,000 to US$15,000 in 1975 US dollars.[14] For an extra 80,000 tonnes of annual output he estimates that capital expenditure would be around US$1.7 billion, not a daunting prospect compared with the sums that would be needed to meet projected increases in the consumption of other metals. For a projected increase in copper consumption Mikesell estimates capital expenditure at US$58 billion, and for nickel US$12.5 billion. All figures exclude expenditures for exploration and pollution abatement. In its recent report the Tin Council does not foresee any problems in raising money for the industry, nor does it believe that reserves, as currently estimated, would be inadequate.

Some forecasts of Malaysian output are certainly disturbing, because if there were serious difficulties in expanding output in the largest producing country, and also a low-cost one, more would be required of other countries. Of these, both Zaire and Nigeria would have to enjoy a major reversal of fortune if they were to make a contribution corresponding to that of, say, the forties or early fifties. Both Australia and Brazil would have to accelerate their growth sharply to reach, for them, unheard-of levels of output. A recent Australian report suggests that Indonesia, Thailand, Brazil and China could achieve sizeable increases, with smaller increases in Australia and the UK, giving a potential increase of about 100,000 tonnes per year by the year 2000.[15] On the basis of recent performance, Thailand emerges clearly as the most promising source of greater output, given reasonably stable conditions. Moreover, one should perhaps be cautious in accepting a pessimistic view of long-run prospects for the Malaysian industry. While it is true that Malaysian output had been disappointing during much of the seventies after seven good years from 1967 to 1973, production increased by as much as 11,500 tonnes, or about 20 per cent, between 1965 and 1968. This was much more than was forecast by the Malaysian reply to the Tin Council's questionnaire in the early sixties, when it was believed that the reply was unduly conservative in its assessment of future potential. Malaysian output of 73,800 tonnes in 1970 was nearly one-third higher than the highest estimate in the reply. Although there has been a considerable run-down of immediately workable reserves for many dredges on their existing land, it might be a mistake to take a pessimistic view of the Malaysian industry's potential under favourable conditions.

If the postwar recovery is excluded, the last big expansion of world tin production occurred in the interwar years and was associated with a substantial fall in production costs arising from technological improvements over a large part of the industry. Future prospects are uncertain. According to Bernard Engel, 'the cheaply recovered alluvial deposits of south-east Asia are unlikely to provide sufficient reserves into the 21st century, and the expensively mined lodes worked by Bolivian and Australian enterprises will not be fully exploited at present price levels – high as these may appear in historical money terms'.[16] On the other hand, in a general review of world minerals, Mikesell comments that 'most mineral economists believe that depletion of world mineral resources is not likely to be a problem in the sense of substantially raising the cost of finding and processing non-fuel minerals during the remainder of the present century'.[17] This conclusion is based on the expectation that the cost-reducing effects of new technology will more or less keep up with the cost-increasing effects of having to exploit lower-grade deposits. Whether and to what extent tin is an exception to this argument is important for the future rate of growth in consumption, given the pressures for economizing in the leading consuming countries. How far the demand forecasts discussed here are likely to be valid is a matter for judgement or guesswork, depending on one's point of view, since the record of commodity forecasting is at best mediocre, not least because forecasters tend to assume a surprise-free future.

Notes

1. A number of commodity models and the modelling process are reviewed by Walter C. Labys in a contribution to the symposium, *Stabilising World Commodity Markets*, edited by F. Gerard Adams and Sonia Klein, Lexington Books, D.C. Heath, Lexington, Mass., 1978. There is a critical analysis of commodity forecasting techniques by John E. Tilton, *Material Substitution: The Experience of Tin-Using Industries*, Penn. State Univ., Univ. Park, Penn., 1980, Ch. Five. Commodity forecasting with reference to the copper industry is discussed by R.F. Mikesell, *The World Copper Industry*, John Hopkins Press for Resources for the Future, Balt., 1980, pp. 172–86. A new approach to econometric modelling for the world tin market has recently been suggested by Hideo Hashimoto, an economist with the Commodities and Export Projections Division of the World Bank. Hashimoto is specifically concerned with decision-making about investment. He suggests that 'the difficulty in modelling investment behaviour lies in its reciprocal relations with market conditions in the future', that 'present investment will affect market conditions [such as prices] in the

future, and prospects for the latter also influence the former', and therefore, 'modelling investment requires solving present and future market conditions simultaneously'. In his view, existing modelling techniques have failed to incorporate this simultaneity into models, and hence do not 'accord with corporate business practices'. The model he presented in a paper to the Fifth World Conference on Tin in Kuala Lumpur, in October 1981, is an attempt to overcome the deficiencies of existing models. See Hashimoto, *A Tin Economy Model for Decision-Making about Investment in Production Capacity*, a paper published by the ITC, London.

2. E.C. Hwa, 'Price Determination in several International Primary Commodity Markets: a Structural Analysis', *IMF Staff Papers*, March 1979, Vol. 26, No. 1. See also Gordon W. Smith, 'US Commodity Policy and the Tin Agreement', in David B.H. Denoon (ed.), *The New International Economic Order: a US Response*, Macmillan, London, 1980, p. 196.

3. Tilton, op. cit., p. 266. Tilton has come to the conclusion that of the three main kinds of forecasting techniques, namely, statistical procedures analysing past trends and projecting them into the future, models specifying causal or behavioural relationships, and techniques that are qualitative or judgmental in nature and less quantitative than the others, only the third seems to 'offer any hope of adequately accounting for the effects of substitution in predicting future mineral requirements, at least over the longer term'. In his opinion, 'this is the only approach that can take full account of the abrupt and inconsistent, yet major, effects of materials substitution that are caused not only by shifts in material prices, but also by changes in technology, government regulations, and other factors as well. See the full analysis in Ch. Five, especially pages 260–9.

4. The Paley Commission's 1975 forecast for the US was 85,000 long tons. In no year of the seventies did US consumption exceed 60,000 metric tons.

5. ITC, *Report on the World Tin Position*, London, 1965. Also World Bank Staff Commodity Paper No. 1, *The World Tin Economy: an Econometric Analysis*, June 1978.

6. World Bank, op. cit., p. 37.

7. Wilfred Malenbaum, *World Demand for Raw Materials in 1985 and 2000*, McGraw Hill, New York, 1978.

8. Malenbaum, op. cit., p. 9.

9. Malenbaum, op. cit., p. 62.

10. Malenbaum, op. cit., p. 17.

11. Malenbaum, op. cit., p. 17.

12. Chase Econometrics, 'Metal Investment in the Eighties', reported in ITC, *Notes on Tin*, April 1980.

13. Yahya and Hollands, op. cit., p. 20.

14. R.F. Mikesell, *New Patterns of World Mineral Development*, p. 95.

15. M.H. Govett and H.A. Robinson, *The World Tin Industry: Supply and Demand*, Australian Mineral Economics, Sydney, New South Wales, 1981.

16. Bernard C. Engel, 'International Tin Agreements', a paper to a seminar at the LSE Centre for International Studies, January 1977, published in Geoffrey Goodwin and James Mayall (eds.), *A New International Commodity Regime*, Croom Helm, London, 1980, p. 91.

17. R.F. Mikesell, op. cit., p. 9.

11 LONG-RUN MARKET PROSPECTS

Government Policy and Tin Mining in Developing Countries

Throughout the seventies there was much discussion of the problems of the world mining industry, especially in developing countries.[1] As a result of uncertainties over the treatment of foreign-owned enterprises in many developing countries, there has been a serious fall in mining investment. A larger proportion of a reduced investment flow has gone to the few advanced countries with a good mineral potential, a more stable and congenial political system, a more certain, if still onerous tax system. Nevertheless, it is still widely believed that the result of this reorientation of new mining investment must be the neglect of better mineral deposits in some developing countries.

As far as tin is concerned, there is little scope for switching investment, since by far the greater part of non-communist world tin reserves is located in a small number of developing countries. This virtually unique position of tin among the large-volume non-ferrous metals makes the attitude of governments in developing countries of crucial importance for the future of the tin mining industry. However satisfactory the reserves position and the future potential of the industry, the key issues are whether and on what terms miners are allowed to extract the tin.

In parts of the industry the question of ownership has been settled. In these countries the prospects for future development now depend chiefly on the state mining enterprises. Their future success, in turn, depends on the resources they are allowed to put into mining and the freedom they have to run the industry. The Bolivian situation is particularly difficult, because the problems of a state mining enterprise in a politically unstable country are superimposed on the physical geography of an industry which makes tin mining far more difficult than in any other country. The so-called Bolivian problem has always been central to the vexed question of the appropriate buffer stock price ranges and to other controversial issues arising from the International Tin Agreement.

At various times since nationalization in 1952, the Bolivian state corporation has received substantial foreign help. While output recovered from the extremely low levels of the early sixties, the lack of growth in the seventies clearly indicated that much more had to be done if the industry were to achieve a significant expansion. A leading Bolivian mining authority lamented in 1979 that the government creamed off too much of COMIBOL's operating surpluses, that the corporation suffered from outdated mining machinery, and from the requirement that all its concentrates had to be sold on unfavourable terms to the state-owned smelting company. In the private sector, still an important part of the industry, the lack of growth was blamed on 'the natural exhaustion of reserves, uncontainable increases in production costs, an insufficient amount of new investment, a lack of exploration, deficient mill recuperation, and other imperfections in both the infrastructure and company management'.[2]

Bolivia, unfortunately, has a highly unstable political and social system, and the mining industry has long been hampered by political and social problems which can be solved only by Bolivians themselves. The government's dependence on mining revenue has been responsible for the persistent tendency to leave the industry too little for investment, although it is debatable whether the tax burden has generally been excessive. Yet the potential of a modernized industry would be a better source of revenue than an industry continually on the margin of bankruptcy. Where the tin-importing countries could help, would be through a major investment programme which paid off in terms of more efficiently produced tin.[3]

From the point of view of importers a good case could be made for such a programme. Firstly, it would reduce the pressure for a higher tin price in the future. A big reduction in tin losses at the mines is long overdue in Bolivia, where new or modernized mines would lead to an increase in output. Secondly, higher productivity would make possible a rise in living standards to something more like those obtaining in other mining industries. There is an enormous difference, for example, between the earning power of a tin miner in Tasmania or Cornwall, and the miners of COMIBOL, both producing an approximately identical high-value non-ferrous metal. It should be possible to make some progress in narrowing this gap, although natural conditions and differing technologies must dictate a large difference in living standards and in economic rent. Thirdly, an investment programme financed largely by OECD governments would accord with the general policy of helping the poorest developing countries in a constructive way. Fourthly, if Bolivia really wants to produce more tin, it seems desirable that the governments of importing countries should put money into the industry. The West should learn from its experience with Libya and the Gulf states that small, thinly populated, wealthy, mineral-rich countries are unlikely to make a contribution to the world's mineral resources commensurate with their resources. Their economic interests do not lie in expansion, since they believe that they have reached the optimum output from the point of view of their own national economies.

In the other producing country where most of the industry is run by a state mining enterprise, Indonesia, there is the problem of reaching a successful compromise on joint ventures between the government and foreign mining companies with the necessary expertise and capital. The conditions which are now laid down by the government are very different from those which expatriate mining companies used to expect. They are illustrated by the history of foreign investment legislation in Indonesia over the last 15 years, during which the government has been more interested in foreign participation in mining than was its earlier post-independence predecessor.[4] After the relatively favourable treatment accorded by the Foreign Capital Investment Law of 1967, the prospective foreign investor was faced with much less attractive third generation contracts, involving the following: a 10 per cent tax on unprocessed minerals exported; the requirement that a foreign company should sell 51 per cent of its shares to Indonesians within ten years of the start of production (compared with only 25 per cent under previous arrangements); a windfall profits tax of 60 per cent when returns on investment averaged over 15 per cent for three years; the requirement that, instead of being allowed free use of their cash reserves, foreign companies should deposit sales-generated foreign exchange in central bank approved accounts, where the money was first converted into rupiahs and then reconverted into foreign exchange, the central bank collecting a commission on the transaction. These provisions so discouraged foreign investors that no new private foreign capital was invested in Indonesia's tin mining industry for some years. In 1980, proposals were at last made to lighten the tax burden on foreign investment projects and modify the restrictive clauses in foreign contracts, including the currency requirements.

The problems arising over land and exploration rights in Malaysia have been referred to in an earlier section. There is some evidence of improved prospects as a result of greater participation by state governments, which according to the Tin Council report will now be more willing to relax their strict policy on making land available for mining. Thus areas previously inaccessible can now be explored for new deposits which may well be mined in the future. There is, however, a complication from the federal government's policy of giving the Malays a larger stake in tin mining. Discussing the problems of relations between host governments and foreign investors, the Tin Council report comments on Malaysian policy:

> where countries insist on a certain percentage of local investment in proposed new mining projects, or where a proportion of the equity of a new mine is reserved for a particular ethnic group within a country, there can be a problem in obtaining the required amount of capital from the appropriate division of the investing community, particularly where the reservation is aimed at the economic improvement of a less privileged section of the community who, perhaps, have less capital to invest.[5]

In general, however, there does not appear to be any problem in obtaining investment money for tin mining, provided projects are believed to be economically viable. The Tin Council has not had any 'report of major projects being held up for lack of investment funds, although in some countries there is some difficulty in obtaining sufficient foreign exchange funds to cover the necessary external costs of a mine, and the political risk of investment in certain countries has been known to weigh the balance against multinational participation in marginally economic projects'.[6] Some projects in uncertain areas may be financed by international organizations. The Tin Council quotes a European Investment Bank loan to the Societé Minière of Rwanda for a mining project and smelter. Another good example is in the long-dormant Burmese industry, which is likely to benefit from improved relations between the government and the West. A World Bank loan of US$16 million was given to the government for the Heinze Basin tin and tungsten project after feasibility studies had been financed by the United Nations Development Programme and the International Development Association.

Technological Change and Consumption

It may be expected that users will continue to look for ways of economizing tin, but there is no reason to suppose that the process will be steady or that research will be all one-way. As the detailed Tilton inquiry into American experience has found, 'while technological change has been a powerful force shaping the use of tin in all the applications examined . . . its impact has often been abrupt and inconsistent, at times stimulating and at other times curtailing its use.'[7] The effect of this random and discontinuous element in the process of innovation is to make it difficult to foresee its results.

There seems to be general agreement that the price of tin has been an important factor in encouraging large users to spend money on ways of reducing consumption. This has been particularly noticeable in the US, where the pressure of government was also felt throughout the forties, compelling and exhorting firms to economize. Before the last war, it is possible that 'inertia in traditional material usage' tended to check substitution, but as Canavan has found, there is now little evidence of inertia retarding substitution. On the contrary, 'material and total soldering costs, technological change, and more recently, public policy, have induced a net decline in tin input for many products, and furthermore, bode a continuation of this decline'.[8]

The continued influence of price has been stressed in a recent major survey of the packaging industry. The Rayner-Harwill study has argued that 'the only likely check to tinplate's future growth in both the food sector and beverage can market is the price of tin, with US$10 a lb. the potential point where a switch to alternative materials might be considered economically justified'.[9] The average monthly price in 1979 was US$7.07, rising in the first half of 1980 to US$8, roughly the same in real terms as in 1979. It is only speculative, however, to specify a particular cut-off price, since new technological developments and changes in the prices of competing materials could raise, or even lower, the point at which users were tempted to make the switch. Moreover, there are usually capital costs to consider in doing so.

The Rayner-Harwill survey, analysing the competition from aluminium and TFS, stated categorically that 'it would appear to be quite certain that tinplate holds the whiphand in terms of future development potential and cost effectiveness'.[10] It based this strong conclusion on several factors, which, in its opinion, indicated the progress made by the steel producers in countering the challenge from aluminium and their continued preference for improvements in the tin coating. Aluminium, it was argued, would not be significant in the huge food market, because it was 'not economic at gauges demanded for good performance'.[11] Tinplate remained the only material that could be used in all can-making processes for all products, with both versatility and reliability. The survey argued that steel-based two-piece cans would increase their share of the beverage market in the US at the expense of aluminium, and that in Europe new two-piece can-making equipment would use steel on grounds of lower costs. An ITC paper to the Fifth World Conference on Tin in October 1981, however, is less optimistic about the future of the important US beverage can market for tinplate. Noting that the number of steel beverage cans in the US had fallen by nearly 20 per cent between 1976 and 1979, and given the prevailing pattern of investment in new aluminiun can-making capacity, the paper foresees that 'tinplate's share of this market will continue to fall',[12] at least up to 1985, the limit of its forecast. The Rayner–Harwill survey did not believe that TFS was capable of significantly increasing its threat to tin in the short run, since both steel producers and can makers still found that the coating with tin was cost-effective. The ITC study is more cautious, particularly about the penetration of parts of the Japanese market by TFS. The Rayner–Harwill survey expected some loss of tin's traditional tinplate markets through the development of still thinner tin coatings and the replacement of soldered by welded and cemented cans.

A problem facing tin in food cans is the growth of opposition to lead on health grounds, which is stimulating governments to impose legal constraints on the use of solder where it comes into contact with food. If public policy leads to the elimination of solder on cans, there will be a minor loss to tin. This would not effect the massive consumption of tin as a coating for the million of tonnes of tinplate. Here the serious problem arises from the growing campaign against non-returnable containers, which has been developing during the seventies in some advanced countries with a high tinplate consumption. The proliferation of aluminium beverage containers has been an important cause of dissatisfaction with non-returnables, but used aluminium cans are more economically recycled than used tinplate cans, and hence less vulnerable to environmental legislation, which tends to hit tinplate harder than aluminium. The view that aluminium is necessarily better than tinplate from the social standpoint has been challenged at the 1980 Tinplate Conference. One defender of tinplate quoted an American inquiry which had apparently established that 'twice as much fuel was being consumed to distribute a case of soft drinks in Oregon as in Washington, chiefly because of the need to make additional trips to collect empty containers'.[13] However, it is difficult to make a comprehensive assessment of relative costs, particularly one which decisively affects the argument over environmental damage.

The effect of environmental legislation on the use of cans is seen from the experience of several US states, which have been pioneering public policy in this respect.[14] As a result of legislation against non-returnables in the state of Oregon, the share of canned beverages fell from 33 per cent in 1971 to only 10 per cent in 1978. In Michigan there was a fall from 54 per cent to 33 per cent between 1977 and 1979. In other states there was stronger opposition to legislation. Within the European Community the Commission has tried to harmonize the different approaches towards packaging, waste management and preservation of the environment.[15] It has suggested that the Community packaging industry should try to work along the following lines: design products for ease of reclamation, recycling and disposal; use the least energy-intensive material that will do the job; reduce certain polluting effects of packaging materials. So far, Denmark's legislation has been the most restrictive. A 1971 Act on beverages led to an agreement with industry to phase out gradually the sale of beer in metal cans by 1981, and later to an administrative order in 1979 which banned the sale of carbonated soft drinks in non-refillable containers. Under a 1977 Act steps are being taken to include beer in the ban on non-refillable containers. At the other extreme, according to the Commision, is the attitude of the UK government, which prefers that industry should work out its own destiny.

How tinplate will be affected in the long run by the spread of health and environmental legislation in developing countries remains to be seen. Tinplate certainly has a strong historical position as a safe packaging material in which a thin tin coating plays a vital function. The contribution to the welfare of the mass of the people in the industrial world can hardly be overestimated, a point which should be borne in mind by the environmentalist lobbies.

In its main use, tinplate, tin is still vulnerable to the extent that steel producers are essentially in business to sell steel. Although the 1980 Tinplate Conference was in many respects reassuring for tinplate, and hence tin, in terms of technological advances for maintaining its position as the leading material for the packaging of heat-processed foods and drinks, there was evidence of the continued pressure on tin. German and Japanese research was going on into a new can stock material, LTS, or lightly-coated steel with a thinner coating. There was also an Australian report of progress on the blackplate can, that is, one with no tin coating, and indications that in Japan tinplate still faced strong competition from TFS.

The US Steel Corporation has reported the invention of 'a cost effective water-based lubricant system which eliminates the need for tin in the production of two-piece steel beverage cans'.[16] This could be either a major breakthrough or just another of the hazards with which tin has been faced over the last few decades. It was noted in the Rayner–Harwill report that, in competing with organic coatings, tin was competitive at US$5.60 a lb., and could go up to US$10 a lb., before it lost to these coatings in the beverage market. At the higher cost, capital expenditure on the switch would be more than offset by the cheaper organic coatings.

There is evidence of continuing substitution in the other uses of tin. In the US, after the fall in the tin content of solder used for passenger automobiles from 23.3 per cent to 3.1 per cent during the war, there was a long period of no change. Since 1975 there has been a slight but possibly significant change. The conclusion has been reached in the study carried out by Canavan that by 1985, so far as the US is concerned, 'no tin-lead solders will be used on mass-produced automobile bodies as filler'.[17] The reasons, according to Canavan, are only partly economic in the sense that manufacturers will find it more profitable to reduce the consumption of solder; more important is likely to be the pressure from the government to cut down the lead hazard. Canavan also detects a possible fall in another automobile consumption of tin:

> All indications point to a further decline in the tin input per radiator in the future. Even if the welded or plastic tank is not adopted, the average tin content of radiator solders is likely to drop, perhaps to less than five per cent compared with the 1978 average of 16 per cent, in response to the escalation of tin prices.[18]

One of the newer, if minor uses of tin has also become vulnerable in recent years. The aerosol can, which became popular in the sixties, reached a peak output of 2,772 million units in the US in 1973. After running foul of environmental problems, production fell by one-quarter in the later seventies and the use of high-tin solders in aerosols fell drastically.

After a period of growth in certain chemicals, there are signs of some erosion of this source of demand for tin, apparently as the result of adverse changes in relative prices. Derek G. Gill's study of American experience suggests that the price of tin has been a major factor behind the attempts to reduce the tin content of stabilizers. In his judgement, cheaper metals such as cadmium, barium and antimony, though still at the testing stage, will possibly replace organotins in the long run. 'The use of antimony stabilizers is expected to grow rapidly in the future, taking all or most of the market now held by stabilizers. In the process, the tin content of stabilizers is likely to fall to zero'.[19]

Tin Consumption in the Centrally-planned Economies

With the growth of their industrial output and *per capita* real incomes in the last two or three decades, the Soviet Union and the other East European members of the COMECON might be expected to have increased their tin consumption substantially. As they narrow the gap between their level of development and that of the European Community, their tin consumption might reasonably approximate to that of the Community, provided material conditions of life become similar. It is, after all, a fundamental justification for their economic system that it is capable of matching, even surpassing, the material achievements of the West.

It could be argued that the COMECON countries are likely at the present time to have a higher consumption of tin per unit of output tin-using manufactures than Western countries, because technological innovation has been less than in the West over a wide range of products. The COMECON countries for example, have lagged behind in electrolytic tinplate. Even now, their average tin coating on tinplate is considerably thicker than in the leading European countries because there has been less investment in electrolytic capacity, which has had a lower priority than other forms of manufacturing investment. It can also be assumed from the nature of the system that there has not been the same competitive pressure in the packaging industry among different materials and methods of production. There is likely to be a similar lack of competition to stimulate tin-economizing technology in other industries.

In so far as this interpretation of the position in COMECON countries is correct, it would be possible in the future to increase industrial output and real income with a disproportionately small increase in tin consumption, provided the latest Western technology were adopted. Beyond a certain point, the relationship between tin consumption and its various determinants would tend to behave as in the West, at least as far as the long-run trend was concerned.

The combined tinplate production of the COMECON countries which are members of the International Tin Agreement was put at 877,000 tonnes in 1978, requiring an estimated 5,800 tonnes of tin. For the purpose of comparison it might be noted that in the same year the UK alone produced 25 per cent more tinplate with 6,000 tonnes of tin. Consumption by the same COMECON countries in 1978 has been put at 1,164,000 tonnes, only 15 per cent more than in the UK with its population about one-sixth of that of the COMECON group. The Soviet Union is estimated to have used only 10 per cent more tinplate than the German Federal Republic in 1979 with a population about four times as large as the Republic. In fact, the Soviet Union consumed only 20 per cent more tinplate than Brazil.

These comparisons show the very great differences in living standards between industrial countries in Eastern and Western Europe, as measured, at least, by the consumption of canned goods, in spite of the former countries' industrial progress since the Second World War. In this respect, the COMECON countries are roughly where some advanced West European countries were nearly 30 years ago. If, therefore, the pattern of consumption in Eastern Europe moves towards the pattern which is common in contemporary Western Europe, there must be a large increase in the production of tinplate and canned goods. In 1978 the European Community used 22,400 tonnes of tin for about 4 million tonnes of tinplate, compared with the 5,800 tonnes of tin used by the COMECON group for its 877,000 tonnes of tinplate. This difference in tin consumption was equal to about 9 per cent of non-communist tin consumption in 1978.

Tin consumption in centrally-planned economies depends on decisions by the planning authorities, affecting the volume and composition of industrial output and the raw materials used for that output. Unless new tinplate and can-making capacity is included in the plan, consumption of canned products can only increase to the extent that imports are allowed to grow. However, consumer tastes and habits must play some part in influencing decisions by the planning authorities, certainly more now than they would have done ten or twenty years ago. There will be a tendency for rising real incomes, at the levels now reached by a large proportion of the population, to be associated with a greater demand for canned goods. Increases in the supply would be a logical response to the pressure for a better standard of living in countries where the population has been making sacrifices for decades.

There are clear signs that the planners have responded to this presure. It can be seen in the growth of tinplate production and consumption in several COMECON countries. Czech consumption of tinplate rose from 32,000 tonnes to 105,000 tonnes between 1965 and 1979, Polish consumption from 43,000 tonnes to 146,000 tonnes, Hungarian consumption from 17,000 tonnes to 53,000 tonnes. Most Czech and

Table 11.1: East-West Trade in Tin Metal, 1960–1980 (000 tonnes)

	To West from China[a]	To East from West		Net trade
		To Eastern Europe	To USSR	(+ to West (– to East)
1960	4.6	–	–	+ 4.6
1961	6.4	1.9	–	+ 4.5
1962	5.5	3.5	0.9	+ 1.1
1963	7.1	2.2	3.3	+ 1.6
1964	6.1	1.1	4.1	+ 0.9
1965	6.1	1.3	5.8	+ 1.0
1966	4.2	0.4	4.8	– 0.6
1967	2.7	1.4	5.7	– 4.4
1968	3.5	1.9	7.1	– 5.5
1969	3.2	1.6	6.8	– 5.2
1970	3.8	1.5	8.3	– 6.0
1971	5.0	1.0	4.4	– 0.4
1972	6.7	6.3	4.3	– 3.9
1973	8.0	8.2	4.0	– 4.2
1974	8.6	9.1	5.2	– 5.7
1975	11.2	6.5	9.7	– 5.0
1976	6.2	8.9	9.8	–12.5
1977	2.9	7.8	9.7	–14.6
1978	5.1	7.6	14.1	–16.6
1979	3.2	7.6	13.7	–18.1
1980 prov.	2.5	6.1	14.2	–17.8

Note: a. West excludes centrally-planned economies.

Source: ITC, *Trade in Tin 1960–1974*; *Tin Statistics*, 1965–79; *Tin International*, various issues.

Polish demand was met from newly installed electrolytic tinplate capacity. The estimated increase in Soviet consumption was 50 per cent between 1967 and 1979.

No official figures for total tin consumption in the COMECON countries are available to the Tin Council, but it has been estimated that Bulgaria, Czechoslovakia, Hungary, Poland and Rumania averaged 9,400 tonnes in 1965-6 and 12,900 tonnes in 1978-9.[20] East-West trade figures, given in Table 11.1, show that imports of tin metal into both the Soviet Union and the other COMECON countries increased considerably between the sixties and the seventies. With the decline in Chinese imports to the West, there was a sharp increase in the net outflow from West to East in the late seventies. By the end of the decade this was a useful support to the market price, with depression in the Western industrial countries and an increase in non-communist production.

As economic development continues in the COMECON countries,

there must be a large increase in the production of tin-consuming goods such as electronic equipment, as well as of canned foods and beverages. The effect on tin consumption will depend on the extent to which advanced technology is used in order to economize in the tin content per unit of output. It will also depend on their general policy towards raw material imports from non-communist countries.

Tin Consumption in the Developing Countries

Tin producers look to developing countries with the majority of the world's population for the long-term growth of demand which will more than offset the slow growth or stagnation of demand in the present-day industrialized countries. The basis for this hope is the great difference in *per capita* consumption of tin between developed and developing countries. There is a range from about 0.5 pounds of tin *per capita* in the former to a mere 0.02 pounds in India. Even in a more advanced developing country like Brazil, with a rapidly growing tinplate industry, it is less than 0.10 pounds. It would seem to follow, therefore, that even allowing for the tin-economizing effects of technological change, the spread of industrialization, and rising living standards among the mass of people in developing countries, must raise tin consumption in the world as a whole.[21]

There has already been a substantial expansion of industrial output among a number of developing countries, especially in the group now usually referred to as the newly industrialized countries. A sizeable number of people in the developing countries have a higher standard of living than in the fifties or sixties. To some extent this has certainly been reflected in the direct and indirect consumption of tin in some developing countries. From the point of view of tin producers, however, this does not entirely represent an increase in tin consumption, since part of the increase is at the expense of tin-using manufactures in the industrial countries, either for domestic consumption or for export. The creation of tinplate capacity, for example, may cut imports of tinplate from industrial countries, while the expansion of exports of electrical and electronic goods may reduce domestic production in industrial countries, as is shown by the experience of both the US and UK.

Although rising real incomes should raise tin consumption in developing countries, a substantial addition to world demand is not likely to come about quickly. There are several reasons why it would be unrealistic to expect more than a very modest growth for many years. Consider the most important use for tin, the packaging of food in tinplate containers. In all developing countries canned food is relatively expensive and may be expected to remain so, for most people in these countries are still at a level of income which is too low for spending much on canned food. The increases in *per capita* incomes which are likely for years to come must still leave the great majority of the population far below the level of income common in the industrial countries.

While the demand for canned food tends to rise with income, it is not only the level of income which determines consumption. Much depends on consumer tastes and habits, which normally change only slowly. In developing countries, a traditional pattern of consumption checks the growth of demand for canned food, just as it did in much of Western Europe until recent times. Some countries, however, had a substantial consumption of tinplate in the past as the result of an export trade in canned products, such as canned pineapples from Malaysia or canned meat from Argentina and Uruguay.

It is not surprising that a traditional pattern of consumption should favour fresh food. Labour services in developing countries are generally cheap. As long as labour is cheap, the convenience appeal of packaged food of any kind, an important selling point in high-income, dear labour countries, does not affect even the small minority of well-off consumers. Moreover, the ready availability of fresh food in towns and villages reduces the need for canned food. Easy access daily to fresh food and no lack of labour for preparation and shopping discourage the sale of canned food. It is no coincidence that the increased participation of women in paid employment in most Western European countries in the last few decades has been associated with a change in shopping habits which favours canned food, processed food of any kind and, more recently, frozen food. Thus, if the demand for tinplate, and hence for tin, is to increase more rapidly than it has done, there must be not only a sufficient rise in income, but also a change in the pattern of consumption in developing countries, analogous to what has been going on in Western Europe and Japan for many years.

To some extent the pattern of consumption will be changed by the growth of the urban population, provided this does not simply mean a massive increase in the urban poor. City dwellers are more exposed to the selling pressures of the market. Traditional tastes and habits are weakened by the switch from country to town living. The effects would be felt not only on canned foods, but on other uses of tinplate, for example, aerosols, which have owed their expansion to the fact that they can perform functions for which there is no convenient substitute.

The demand for tinplate in developing countries may also be affected by the domestic production of electrolytic tinplate. If there is no domestic industry or inadequate capacity, a shortage of tinplate may not be made good from imports because of a shortage of foreign currency. If governments give a higher priority to tinplate capacity in their economic planning, there may be a stimulus to potential users. But in this respect it must be emphasized that the developing of a market for canned foods depends on much more than the supply of electrolytic tinplate. It is possible that locally-produced tinplate may be more expensive than imports, even allowing for a tariff on imported tinplate. Economies of scale in this part of the steel industry may be lost in a developing country. There has to be investment in agriculture, in can-making and canning facilities, and also in transport. The food delivered to the canners must be of a higher and more consistent quality than fresh food for traditional markets. The transport system must be adequate to bring the unprocessed food to the canneries, which in turn must have an adequate system of distribution. In short, unless the whole chain of production and distribution from the supply of steel to the final seller is efficiently organized, the cost of production will check the expansion of consumption. It is also necessary that the scale of production at the different stages must be large enough to keep down unit costs, including overheads. A large number of small plants will lose economies of scale, although they may cut transport costs. Larger plants may have to work too far below optimum rates. Thus if costs are too high for the finished products, the growth of demand will be inhibited.

Some developing countries already have substantial export markets for canned food. It is likely that under favourable circumstances the demand for exports of canned food would have good growth prospects, increasing their consumption of tinplate and eventually making economic the establishment of their own tinplate production facilities. It appears, however, that the expansion of tinplate production in the leading developing countries has been due primarily to domestic demand for tinplate. This is true of Brazil, Mexico and Venezuela, where the growth of GNP per head and of total GNP has been relatively high throughout the sixties and seventies.

Several developing countries with the largest populations are still far down the tin consumption league. The most important are India and Indonesia among the non-communist countries. Tin producers may well look enviously at the potential of the Indian market, remote as it may still be, where even a small can use per person per annum might open up a sizeable new market for tin. Still more enviously, they may look at the People's Republic of China, comparing its potential with the reality of Taiwan, where tin consumption has increased considerably in the last ten years. The growth of Indian consumption of tinplate has been extremely slow, illustrating the combined deterrent effects of low average incomes, conservative consumer habits, and possibly restrictive government policy. Indian production of tinplate averaged over 50,000 tonnes as far back as 1935-7; in 1963 it was 100,500 tonnes rising only to 120,000 tonnes in 1979. Consumption in 1978 was 258,000 tonnes, compared with 142,000 tonnes in 1963. There are indications of a possibly substantial higher rate of consumption following an increase in electrolytic capacity. Total hot-dipped and electrolytic capacity was 377,000 tonnes at the end of 1977. Not much increase in tin consumption, however, is likely in the next few years, since hot-dipped capacity is being replaced with electrolytic. Consumption of tin in 1979 for tinplate was less than in the early sixties.

New electrolytic capacity is being added almost yearly in the developing countries as a whole, but most seem to have some spare capacity. So far, of the tin producing developing countries, only Thailand produces some tinplate. Two other countries, Malaysia and Indonesia, are likely to have capacity in the eighties, since there is a natural desire to consume more locally-produced tin, but account has to be taken of the fact that tin is only a very minor component in tinplate, and investment in tinplate capacity is not necessarily the best way of using scarce capital and skilled labour.

The general picture of the tinplate situation in developing countries does not suggest any dramatic change in tin consumption over at least the next decade. Not much increase has occurred in the last ten or fifteen years. David Lim comes to the same conclusion in a study of the case for local tinplate production by tin producing countries. In his view, 'the consumption of tinplate and hence the use of tin is not going to increase significantly in less developed countries without very rapid growth over a sustained period',[22] a prospect for which the omens at the beginning of the eighties are not promising.

Consumption of tin in other uses has not grown dramatically. In Latin America and Asia (excluding Japan) there was an increase of 3,200 tonnes between 1968 and 1978. Total consumption for all uses, including tinplate, in these areas rose from 14,900 tonnes to 21,000 tonnes, with most of the increase coming in a small number of newly industrializing countries, the largest of which is Brazil. Two small developing countries which have achieved very high rates of growth in the last 10 to 15 years, South Korea and Taiwan, have greatly increased their tin consumption. It is an indication of the rapid growth of their metal-using production that tin consumption grew nearly sixfold in ten years. Both now use more tin than Sweden, Switzerland or Denmark. As major exporters of electrical and electronic equipment to the West, South Korea and Taiwan are probably displacing some tin consumption in the West, but there is also a considerable local consumption of tin-using products.

Not much information is available about the breakdown of non-tinplate uses, but much of it probably takes the form of solder. It is likely that this is largely primary tin solder, in contrast to the high percentage of secondary tin used in some Western countries, where there should be a large stock of secondary tin in solder as a result of past high outputs of metal-using products with a tin content. No doubt, as tin consumption increases in these newly industrializing countries, the flow of secondary tin in solder and other alloys will increase, but one would expect that for some time the greater demand would be for primary tin.

As the largest of the industrializing developing countries, Brazil seems to promise the largest increase in tin consumption by the end of the century. Between 1968 and 1978 consumption rose from 2,200 tonnes to 5,800 tonnes, making Brazil the eighth largest consumer. Brazilian tinplate production exceeded half a million tonnes in 1978, making it the sixth largest producer. Installed tinplate capacity is about 610,000 tonnes, with the prospect of reaching one million tonnes during the eighties. Brazil, in fact, seems to be the exception to the conclusion reached by Lim that the 'consumption of non-tinplate goods in the tin-producing less developed countries, being strongly linked to the level of economic development' is so low that the 'prospects for producing them for local consumption are not very promising'.[23]

Notes

1. There is a good concise account of the issues arising in this section by R.F. Mikesell, *New Patterns of World Mineral Development*, British-North American Committee, London, 1979. See also L.T. Wells and D.N. Smith, *Negotiating Third World Agreements: Promises as Prologues*, Ballinger, Mass., 1975, and R. Bosson and B. Varon, *The Mining Industry and the Developing Countries*, Oxford University Press, for the World Bank, 1977.
2. *Tin International*, October 1979, p. 392.

3. The World Bank is reported to have agreed to finance the first phase of an investigation into COMIBOL in order to prepare the way for major improvements in the running of the state mining corporation. This is in response to agitation about the problems of the industry, whose total output in 1979 was the lowest since 1967. See *Tin International*, July 1980, and subsequent issues for reports on the progress of discussions on reforms in the corporation and in the taxation system for mining.

4. See the discussion of Indonesian policy towards foreign investment in the *Mining Journal*, 11 July 1980, and also the ITC publication, *Notes on Tin* January 1980, quoting the *Far Eastern Economic Review*, 1 February 1980.

5. ITC, *Tin Production and Investment*, p. 144.

6. ITC, op. cit., p. 143.

7. Tilton, *Materials Substitution: The Experience of Tin-Using Industries*, p. 260.

8. Canavan, in Tilton, op. cit., p. 181.

9. 'The Resurgence of the Tin Can, a summary of the Rayner–Harwill survey, An Appraisal of Tinplate and Aluminium in the Packaging Industry', *Tin International*, January 1979.

10. *Tin International*, January 1979.

11. Ibid.

12. U. Yahya and N.C.P. Hollands, International Tin Council Tin Industry Officers, 'World Tin Consumption: Present Position and Medium-Term Forecast', paper to the Fifth World Conference on Tin, Kuala Lumpur, October 1981, p. 10.

13. *Tin International*, November 1980, reporting on the Second International Tinplate Conference, London, October 1980.

14. Tilton, op. cit., p. 60.

15. See *Europe 81*, Nos. 1/2, the magazine of the UK Offices of the European Commission, p. 8.

16. *Tin International*, October 1980.

17. Canavan, in Tilton, op. cit., p. 162.

18. Canavan, p. 142.

19. Derek G. Gill, 'Tin Stabilisers and the Pipe Industry', in Tilton, op. cit., p. 230.

20. East German tin production and consumption in 1979 have been put at 1,600 tonnes and 3,000 tonnes, respectively, by Minemet; see *Minemet Annual*, 1979, quoted in *Tin International*, March 1981.

21. The tinplate market in developing countries is discussed by W. Robertson, 'The Market for Tinplate in Developing Countries', a paper to the London Conference on Tin Consumption, 1972, published in the *Proceedings*, ITC, London, 1972.

22. David Lim, 'Industrial processing and location: a study of tin', *World Development*, Vol. 8, No. 3, 1980, p. 210.

23. Lim, op. cit., p. 211.

BIBLIOGRAPHY

Adams, F.G. and Behrman, Jere, R., *Econometric Modelling of World Commodity Policy*, Lexington Books, D.C. Heath, Lexington, Mass., 1977

——— and Klein, Sonia A., *Stabilizing World Commodity Markets*, Lexington Books, D.C. Heath, Lexington, Mass., 1978

Banks, Ferdinand E., *The World Copper Market: an Economic Analysis*, Ballinger, Cam., Mass., 1974

——— *The Economics of Natural Resources*, Plenum, New York, 1974

Barkman, Kerstin, 'The International Tin Agreements', *Journal of World Trade Law*, Vol. 9, 1975

Barry, B.T.K., 'Tinplate 1978 – an International Perspective', a paper to an Australian Tinplate Conference, Sydney, 1978

Barry, Thomas B., 'The Competitive Challenge to Tin', *Tin International*, Nov. 1978

——— and Littler, Dale, 'Competition in the Can-making Industry', *Tin International*, Oct. 1979

——— and Littler, Dale, 'Two-piece Can-making: an Overrated Technology?', *Tin International*, May 1979

Barton, D.B., *A History of Tin Mining and Smelting in Cornwall*, D.B. Barton, Truro, England, 1967

Bosson, R. and Varon, B., *The Mining Industry and the Developing Countries*, Oxford Univ. Press for the World Bank, 1977

Brown, C.P., *The Political and Social Economy of Commodity Control*, Macmillan, London, 1980

Carman, John S., *Obstacles to Mineral Development: a Pragmatic View*, edited by Bension Varon, Pergamon Policy Studies, New York, 1979

Chase Econometrics Associates, *Metal Investment in the Eighties*, New York, 1980

Cuddy, J.D.A., *International Price Indexation*, Saxon House, D.C. Heath, and Lexington Books, Mass., 1976

Cullen, P.W. and McDonald, G.C.R., 'Problems resulting from the development of a large-capacity mineral recovery dredge', in ITC, *Proceedings of the Fourth World Conference on Tin*, Vol. 2, 1974

Denoon, David, B.H. (ed.), *The New International Economic Order: a U.S. Response*, Harvard Univ., New York, 1980

Denyer, J.E., 'The Production of Tin', in ITC, *Proceedings of the Conference on Tin Consumption*, London, 1972

Elliott, William Y. *et al., International Control in the Non-Ferrous Metals*, Harvard Univ., New York, 1937

Engel, Bernard C., 'The International Tin Agreement', a paper presented to a seminar on the 'New International Economic Order and the Commodity Markets', London School of Economics, 1977

Fox, David J., 'The Bolivian Tin Mining Industry: Some Geographical and Economic Problems', in ITC, *Proceedings of a Technical Conference on Tin*, London, 1967

—— *Tin and the Bolivian Economy*, Latin American Publications Fund, London, 1970

Fox, William, *Tin: the Working of a Commodity Agreement*, Mining Journal Books, London, 1974

—— 'Some Thoughts on U.S. Stockpile Disposals', *Tin International*, August 1973

—— 'The Reserves and Availability of Tin and the World Consumer', in ITC, *Proceedings of the Conference on Tin Consumption*, London, 1972

Geer, Thomas, 'The Post-war Tin Agreement', World Bank Paper, 1969

Gilbert, Christopher, *The Post-war Tin Agreements and their Implications for Copper*, Commodities Research Unit, London, 1976

Gillis, Malcolm *et al., Taxation and Mining: Non-Fuel Minerals in Bolivia and other Countries*, Ballinger, Cam., Mass., 1978

Govett, M.H. and Robinson, H.A., *The World Tin Industry: Supply and Demand*, Sydney, New South Wales, 1981

Hallwood, Paul, *Stabilization of International Commodity Markets*, Jai Press, Greenwich, Conn., 1979

Harrington, Peter, 'Nigerian tin mining: better hopes for the future?', *Tin International*, Dec. 1973

Hayes, J.P. (ed.), *Terms of Trade Policy for Primary Commodities*, Commonwealth Economic Papers No. 4, Commonwealth Secretariat, London, 1975

Hedges, E.S., *Tin in Social and Economic History*, Edward Arnold, London, 1964

Herfindahl, Orris C., *Copper Costs and Prices 1870–1957, Resources for the Future*, John Hopkins Press, Balt., 1959

Hoare, W.E., 'Trends in Tin Consumption: Some Technological Observations', in ITC, *Proceedings of the Conference on Tin Consumption*, London, 1972

Hoong, Yip Yat, *The Development of the Tin Mining Industry of*

Malaya, Univ. of Malaya Press, Kuala Lumpur and Singapore, 1969
Hosking, K.F.G., 'The Search for deposits from which tin may be profitably recovered now and in the foreseeable future', in ITC, *Proceedings of the Fourth World Tin Conference*, Vol. 1, 1974
Hughes, Helen, 'Economic rents, the distribution of gains from mineral exploitation, and mineral development policy', *World Development*, Vol. 3, 1975
—— and Singh, Shamsher, 'Economic rent: incidence in selected metals minerals', *Resources Policy*, Vol. 4, No. 2, 1978
Hwa, E.C., *Price determination in several international primary commodity markets: a structural analysis*, IMF Staff Papers, March 1979
International Tin Council, *Annual Reports*
—— *Monthly Statistical Bulletin*
—— *Notes on Tin*
—— *Statistical Yearbook*, 1962-1968
—— *Trade in Tin 1960–1974*
—— *Tin Production and Investment*
—— *Prospects for World Tin Consumption 1957–68*
—— *Prospects for World Tin Consumption up to 1975*
—— *Report on the World Tin Position with Projections for 1965 and 1970*
—— *Proceedings of the Technical Conference on Tin*, London, 1967
——*Proceedings of the Second Technical Conference on Tin*, Bangkok, 1969
—— *Proceedings of the Conference on Tin Consumption*, London, 1972
—— *Proceedings of the Fourth World Conference on Tin*, Kuala Lumpur, 1974
—— *United Kingdom: Tin in Tinplate*
International Tin Research Council, *Proceedings of the First International Tinplate Conference*, London, 1976
Knights, Ian A., 'Deep-water tin mining', *Tin International*, August 1980
Knorr, Klaus E., *Tin under Control*, Food Research Institute, Stanford Univ., Stanford, Calif., 1945
Jackson, F.K.J., 'Changing patterns in Malaysian mining', *Tin International*, June 1980
Labys, Walter C., *Market Structure, Bargaining Power, and Resource Price Formation*, Lexington Books, D.C. Heath, Lexington, Mass., 1980
Lim, David, *Economic Growth and Development in West Malaysia*

1947–70, East Asian Social Science Monographs, Oxford Univ. Press, 1973

—— 'Industrial processing and location: a study of tin', *World Development*, Vol. 8, No. 3, 1980

Malenbaum, W. (ed.), *Material Requirements in the U.S. and Abroad in the year 2000*, Univ. of Pennsylvania, Philadelphia, 1973

—— *World Demand for Raw Materials in 1985 and 2000*, McGraw Hill, New York, 1978

Mikdashi, Z., *International Politics of Natural Resources*, Cornell Univ. Press, Ithaca and London, 1976

Mikesell, Raymond F., *New Patterns of World Mineral Development*, British–North American Committee, London, 1979

—— *The World Copper Industry: Structure and Economic Analysis*, John Hopkins Press, Resources for the Future, Baltimore, 1979

Mills, Rowena, *Statistical and Economic Review of the U.K. Packaging Industry 1974–78: Outlook 1979–80*, Pira, Leatherhead, Surrey, 1979

Minchinton, Walter E., *The British Tinplate Industry*, Clarendon Press, Oxford, 1957

Paley, William S., *Resources for Freedom, Materials Policy Commission of the President of the United States*, Washington, DC, 1952

Prain, Sir Ronald, *Copper*, Mining Journal Books, London, 1975

Robertson, William, *The Tin Experiment in Commodity Market Stabilisation* Oxford Economic Papers, Vol. 12, 1960

—— 'The Market for Tinplate in Developing Countries, with particular reference to Latin America and Asia', in ITC, *Proceedings of the Conference on Tin Consumption*, London, 1972

—— 'The International Tin Agreement', in House of Lords Select Committee on Commodity Prices, Minutes of Proceedings, Vol. 2, 1977

Rogers, Christopher, *The Economics of the International Tin Agreements*, M.A. thesis, Univ. of Liverpool, 1968

Rowe, J.W.F., *Primary Commodities in International Trade*, Cambridge Univ. Press, 1965

Sainsbury, C.L., 'Tin Resources of the World', *Geological Survey Bulletin 1301*, Geological Survey, US Department of the Interior, Washington, DC, 1969

Sauvant, Karl P. and Hasenpflug, *The New International Economic Order: Confrontation or Cooperation between North and South*, Wilton House Publications, London, 1977

Schatz, Ludwig, *The Nigerian Tin Industry*, Nigerian Institute of Social and Economic Research, Ibadan, 1971

Schmitz, Christopher J., *World Non-Ferrous Metal Production and Prices 1700–1976*, Frank Cass, London, 1979

Scholla, Paul F. and associates, *Tin Deposits of Thailand*, Bangkok, 1977

Smith, G.W. and Schink, G.R., 'The International Tin Agreement: an Assessment', *Economic Journal*, Dec. 1976

Spada, A. La, *The World Tin Economy and New Trends*, ITC, London, 1981

Thoburn, John, *Primary Commodity Exports and Economic Development: Theory, Evidence, and a Study of Malaysia*, John Wiley and Sons, London, 1977

——— 'Commodity Prices and Appropriate Technology: Some Lessons from Tin Mining', *Journal of Development Studies*, Jan. 1978

——— 'High prices favour small producers', *Tin International*, March 1978

——— *Malaysia's tin supply problems*, Resources Policy, March 1978

——— *Multinationals, Mining and Development*, Gower, Aldershot, 1981

Tilton, John E., *The Future of Non-Fuel Minerals*, Brookings Institution, Washington, DC, 1977

US Bureau of Mines, *Minerals in the U.S. Economy: Ten-Year Supply-Demand Profiles for Non-Fuel Mineral Commodities 1968–77*, Pittsburg, Penn., 1979

Wells, L.T. and Smith, D.N., *Negotiating Third World Agreements: Promises as Prologues*, Ballinger, Cam., Mass., 1975

World Bank Commodity Paper No. One, *The World Tin Economy: an Econometric Analysis*, Economic Analysis and Projections Department, World Bank, Washington, DC, June 1978

——— Staff Working Paper No. 354, *Development Problems of Mineral-Exporting Countries*, Public and Private Finance Division, Development Economics Department, Washington, DC, August 1979

Yahya, U. and Hollands, N.C.P., *World Tin Consumption: Present Position and Medium-Term Forecast*, ITC, London, 1981

Periodicals

American Metal Market (weekly)
Business Times (weekly)
Latin American Commodities Report (weekly)

Metal Bulletin (weekly)
Mining Journal (weekly)
Mining Magazine (monthly)
Tin International (monthly)

INDEX